The Informed Gardener

The INFORMED GARDENER

LINDA CHALKER-SCOTT

University of Washington Press ❧ *Seattle & London*

THIS PUBLICATION WAS MADE POSSIBLE IN PART
BY A GENEROUS GRANT FROM THE MILLER FOUNDATION.

Printed in the United States of America
Design by Ashley Saleeba
15 14 13 12 11 10 09 08   10 9 8 7 6 5 4 3 2 1

UNIVERSITY OF WASHINGTON PRESS
P.O. Box 50096 Seattle, WA 98145, U.S.A.
www.washington.edu/uwpress

The paper used in this publication is acid-free and 90 percent recycled from at least 50 percent post-consumer waste. It meets the minimum requirements of American National Standard for Information Sciences—Permanence of Paper for Printed Library Materials, ANSI Z39.48–1984.

LIBRARY OF CONGRESS
CATALOGING-IN-PUBLICATION DATA
Chalker-Scott, Linda.
The informed gardener / Linda Chalker-Scott.
p. cm.
Includes index.
ISBN 978-0-295-98790-3 (pbk.)
1. Landscape gardening. I. Title.
SB472.C34 2008
635—dc22   2007047682

Cover illustration: Sarah Dixon, *Summer's Bounty II* (detail), 2004. Oil on canvas, 24 x 48 in. Reproduced by permission from the artist.

This book would not have been developed without the efforts of WSU Master Gardener George Roper. George approached me in 2004, urging me to bring some of my on-line myths into book form, and volunteered to select thirty-five of his favorite columns for this first book. Another WSU Master Gardener, Helen Dowsett, provided the gorgeous illustrations. It is to these passionate WSU volunteers and others who want to know the science behind the myths that this book is dedicated.

CONTENTS

HOW/WHAT/WHEN/WHERE TO PLANT

SOIL ADDITIVES

PREFACE

I didn't start out as an urban horticulturist. My early dream was to be a marine biologist, and I followed this pathway through my first two degrees. As I faced a career crossroad in the early 1980s, my husband suggested I pursue a PhD in horticulture "because I had always enjoyed growing plants."

I followed a fairly traditional laboratory-based program during my doctoral work, eschewing the horticulture classes for the loftier courses in biochemistry and physiology. After all, I reasoned, how hard could it be to plant trees and shrubs? This superior attitude toward practical horticultural science followed me until I moved back home to Washington State in 1997. At the University of Washington's Center for Urban Horticulture, I found myself having to teach the same practical courses I had managed to avoid.

Fortunately, my interest in plant stress physiology was a complementary match for my course teaching. I began to really notice urban landscapes and wonder why trees and shrubs seemed to do so poorly in a climate that should grow magnificent specimens. Why is the average life of a street tree in Seattle only eight years?

Lo and behold! I discovered it wasn't that easy planting and managing trees and shrubs, especially those in urban environments. Many of the recommended practices in soil preparation, plant installation, and landscape management seemed better suited for a cornfield than for a permanent, ornamental landscape—and, indeed, these practices were derived from traditional production horticulture. It is time for a paradigm shift—we need to look at forest ecosystems rather than annual crop fields as the management model for our home landscapes, urban forests, and community greenspaces. As in a forest ecosystem, the mainstays of our landscape are trees and shrubs that are not harvested annually but should live for many decades. The idea for addressing common horticultural myths in a monthly column was born.

This book is a collection of columns written since 2000, initially for use by professionals in the nursery and landscape industry of Washington State. In response to many requests, I began an on-line version of the columns to increase their accessibility outside Washington State. Six years later, the columns are being used by university faculty and students, Extension agents, nursery and landscape professionals, Master Gardeners, landscape architects, and members of the ever-increasing gardening public. The columns generate hundreds of e-mails nationally and internationally. Although the Internet is fast and convenient, many people have asked that I publish these columns in book form.

My intention is not to point fingers but to raise consciousness about a number of misconceptions regarding the management of landscapes dominated by woody plants, or trees and shrubs. As a

PhD student in horticulture, I myself made many of the mistakes discussed in this book because alternative information wasn't readily available and because tradition is a powerful force. I hope to help you avoid or correct these mistakes.

What will you get from this book?

1 You will get information that is science-based. There is a great deal of horticultural information available through scientific journals, but these resources can be difficult to find and even more difficult for a nonscientist to comprehend. This book is based on basic and applied science from university research, originally published in peer-reviewed journals, that has been compiled and presented in a readily understandable manner. The book has been published independently by the University of Washington Press and was not influenced by commercial practices or products. There is no vested interest in anything but the best practices based on currently available science. At the end of most chapters, I've included a short list of current references that are illustrative of those I read before writing about the topic.

2 You will learn to approach marketing claims with objectivity. Interest in gardening continues to grow, and new products spring up like weeds. Some gardening books and Web sites are nothing more than thinly veiled advertisements for products and services. In reality, plants managed to survive long before we (and our hundreds of miracle products) appeared on earth. This book will help you develop a sense of how garden and landscape plants respond to their environment, and in understanding these responses, you will learn how to interpret marketing claims.

3. You will save time. By recognizing and using natural processes in managing your landscape, you will spend less time overdoing soil preparation, weeding, pruning and staking improperly, and replacing plants that have died before their time. Instead, you will spend your time analyzing site conditions, selecting healthy plants, properly preparing the roots for installation, and providing appropriate aftercare. Properly installed trees and shrubs will not be high-maintenance nightmares but instead will allow you to spend your time enjoying the benefits of your landscape.

4. You will save money. Gardening should be a joy, not a financial drain. A landscape planting can be thought of as an investment of your time and money in developing a high-value amenity on your property. Properly installed and managed trees and shrubs have longer lives and seldom need replacing—and the larger a tree becomes, the higher its value. Furthermore, by avoiding worthless or even harmful garden products, you will not only save money but will also produce a healthier, longer-lived, higher-value landscape.

5. You will appreciate your landscape or garden plants as living components. Like people, plants respond to their environment. Light, temperature, water, soil structure and chemistry, microbes, animals, other plants, and, most especially, people will all influence how successfully plants will establish, survive, and thrive. Too often landscape plants are seen only as design elements: While much time is spent choosing the perfect flower color or leaf shape, little thought is given to how plants grow, change, and eventually die. Much of a tree or shrub's survival is dependent on underground conditions and processes—yet these topics are either ignored or misinterpreted in many gardening books.

6 You will reduce your use of fertilizers and pesticides. Indiscriminate addition of chemicals to gardens and landscapes will not solve underlying plant health problems. Many times we assume that a pest or disease is the source of a plant's health problems. When we see symptoms of fungal disease or pest insects, our first impulse is to spray the dickens out of the plant. Unfortunately, this behavior only treats the symptom, not the cause of the original stress. The fact is that most landscape trees and shrubs are killed through poor management techniques. Plants that are under environmental stress are more susceptible to opportunistic diseases and pests. This book will teach you to recognize some of the common stresses associated with landscape plants and give you straight advice on what to do. Not only will you save time and money but you will also protect the health of your family and pets, your landscape, and your surrounding environment by not using unnecessary or excessive chemicals.

7 You will feel better mentally, physically, and spiritually. The information in this book will give you confidence that you are managing your garden or landscape in an environmentally sustainable way, using a forest ecosystem as a model. No more breaking your back with soil preparation and plant management practices better suited for an agricultural field than an ornamental landscape. Instead of expecting instantaneous results, you will rediscover the value of patience in watching a healthy landscape develop. Thus the time you spend in your landscape will become more peaceful as you find yourself becoming part of a sustainable, natural system rather than its adversary.

The Informed Gardener

CRITICAL THINKING

THE MYTH OF ABSOLUTE SCIENCE

The Myth

"If it's published, it must be true."

On several occasions I've been asked how to differentiate between "good" and "bad" science. It's an excellent question, and I'll illustrate it with the review of a book published thirty years ago. *The Sound of Music and Plants* has been cited by dozens of Web sites as solid scientific evidence that classical music benefits plant growth, while acid rock music has a negative effect. This effect is even extrapolated to humans on Web sites with titles such as "How Music Affects Your Kids. . . . What Parents Need to Know" and "Why 'Good Vibrations' May Be Bad." Ignoring the questionable logic of equating plants with humans, an educated nonscientist should be able to determine the validity of a publication. Here's an example of how to approach the process.

The Reality

The Sound of Music and Plants was written by Dorothy Retallack and published by DeVorss & Co. in 1973. The book charts the author's undergraduate research experience at Temple Buell College (now Colorado Women's College) under the direction of her biology professor. The book is a blend of science, music, philosophy, and religion. The actual description of the experiments is relegated to the second appendix. Upon careful reading of the book, I had a number of concerns:

- No scientific rationale or hypothesis is presented; rather, the author exclaims, "What in the world can I do with music and plants!" (Her field of study was music; she was required to take a biology course for her degree.)

- Authorities in unrelated fields are cited to give the appearance of legitimacy to the experiments. The author cites the works of several professors and/or doctors, who are experts in fields such as physics and theology, but certainly not in the biological sciences.

- Various claims are footnoted and referenced, but out of the forty footnotes, only two are both relevant to the subject of plant growth and sound (not music) and from valid scientific sources. The others are irrelevant to the subject matter or are not scientifically valid sources.

- The author anthropomorphizes; in other words, she compares plants to humans in terms of having "likes and dislikes, their feelings and idiosyncrasies." This is poor reasoning and biases her expectations.

- The author claims that "beyond a doubt the phenomenon itself [the effect of music on plants] has been proven." Science does not "prove" any hypotheses: it either disproves or supports a set of assumptions. This is why science is constantly changing as old hypotheses are discarded or amended as we learn more about the natural world.

- The number of replicates is small (four) and probably not sufficient for statistical analysis. Generally, greenhouse experiments would have at least ten replicates for each treatment to quell the "noise" that occurs due to environmental variability. Furthermore, no statistics other than averages are provided in the book, suggesting that statistical variation (the "noise") was not analyzed, and rendering the averages irrelevant. No statistically valid information is provided in the book.

- The experimental design is poor and does not maintain other factors (such as water, humidity, light, et cetera) at optimal and consistent levels. For instance, the potting containers were Styrofoam drinking cups with no drainage, and watering needs were "determined by touching the soil with a finger."

- The book is published by a company that specializes in New Age literature, not science.

- The author did not repeat her experiment nor publish her results in a peer-reviewed journal, nor has any other repetition of the work appeared in this body of literature.

The Bottom Line

Here are some criteria to keep in mind when separating science from pseudoscience:

- If the information is in a magazine or journal, is it a peer-reviewed publication? "Peer review" means that independent experts in the field read and critique the manuscript. By reading a journal's instructions for manuscript submission, it is fairly easy to determine if peer review is part of the process. Lack of peer review in a publication should cause the reader to look elsewhere for scientific verification of the claims in question.

- If the information is contained in a book, who is the publisher? What types of books does the publisher produce? Valid science is usually published by academic or scientific publishing houses, including university presses. Pseudoscience is commonly published by companies with no ties to mainstream science or academia, including vanity presses (i.e., self-publishers).

- Has the experiment been repeated elsewhere? When controversial subjects appear in the scientific literature (e.g., cloning), independent researchers will repeat the experiment to verify the original research. Lack of subsequent scientific verification is a red flag; it means that no one else was able to get the same results.

- A hypothesis becomes a scientific "truth" if repeated scientific experimentation has failed to disprove it. The "laws of physics" are so termed because they have been exhaustively tested, and have not yet been shown to be false. An idea must withstand repeated attempts to disprove it.

- Does the author have an ulterior motive? For instance, is the author attempting to sell a product? Or is the author attempting to sway your thinking on an issue unrelated to science (such as religious morality, as in the book cited above)? Scientists attempt to report their results as objectively as possible.

References

Retallack, D. 1973. *The Sound of Music and Plants.* Santa Monica, CA: DeVorss & Co.

Original article posted in November 2003.

THE MYTH OF INDISPUTABLE INFORMATION

The Myth

"Nursery brochures are always the best sources of information on appropriate planting practices."

As a service to their clientele, retail nurseries often provide educational brochures describing various aspects of landscape plant selection and management. The most potentially useful of these are the instructions for proper installation and maintenance of new plant material. When written lucidly and illustrated carefully, these brochures can maximize plant survival, customer satisfaction, and nursery profits. If employees are also reading these materials, even the least-experienced staff member should be able to answer questions in a knowledgeable manner.

The Reality

Most nursery customers assume that nurseries have adequate staffing and time to read up on the latest horticultural research. This simply is not possible, given the demands of running a business, with the addition of serious nursery problems, such as sudden oak death. Thus, while planting brochures can indisputably be a valuable resource, they can end up causing more harm than good when they are not updated on a regular basis. Like any other branch of science, horticulture is a dynamic field in which experimental science continuously shapes practical application. To be considered as reliable guidelines for good planting practices, resources must be frequently reviewed and revised.

I recently received a planting brochure copyrighted in 1987 that is still being distributed by a retail nursery. Test your knowledge of good planting practices by evaluating the statements from this brochure excerpted below:

- "Dig a hole twice as wide and twice as deep as the plant's root ball."
- "Mix excavated soil with sand, peat moss, rotted manure, or other soil amendments. Note: If your soil is heavy and full of clay, add sand to aid excess water drainage."
- "Gently tap the sides of the container and slip the plant out, being very careful to keep the root ball intact."
- "It's best to water plants in the morning, especially on clear, hot days. If the leaves are allowed to heat up on a hot day and are then splashed with cold water, the plant is shocked and the leaves may shrivel up or spot."

- Bonus point: In general, what should alert you to question the validity of this information?

Here are the most currently accepted planting methods, all of which are detailed in upcoming chapters:

- The hole should be at least twice as wide, but no deeper than the root mass.
- Amending native soil prior to installing permanent landscapes (i.e., woody plant material) is not a sustainable practice; instead, top-dress with organic mulch.
- Container plants should be bare-rooted at installation to remove potting media and to correct root problems.
- Wet foliage is not susceptible to sunburn; although it is best to irrigate in the morning, plants should be watered any time they exhibit drought stress.
- Bonus point: Any planting instructions that were written in 1987 are no longer valid. The science behind the practices has advanced, and educational materials should reflect this.

While it is encouraging that more retail nurseries are providing educational materials to their clients, dated information, whether given orally or in writing, defeats the purpose of this educational opportunity. University Extension faculty and staff recognize the time and money constraints on businesses and are willing educational partners. Those in the nursery and landscape industry should request the most current information on plant selection and management to share with their customers, their colleagues, and their employees. It's information that is readily available from university Extension faculty and staff.

The Bottom Line

- Good planting practices are constantly changing as a result of ongoing scientific research.

- Educational materials for both customers and employees need to be updated annually to ensure validity.

- Most nurseries do not have resources to research and write educational materials.

- Information on good planting practices is readily available from Washington State University and other land-grant university Extension offices.

Original article posted in August 2004.

THE MYTH OF ORGANIC SUPERIORITY, PART 1

The Myth

"Organic products are safer than chemicals."

Recently I received an e-mail from an Internet reader who took issue with my column on compost tea. Among his comments was the following statement: "You talk about groundwater pollution and eutrophication of the watershed from overuse [of compost tea]. Yet, I don't know of any farmers that could afford to overuse the stuff. You don't mention that this kind of pollution results almost every time someone uses petrochemical salt fertilizers. It almost never happens when someone uses compost tea."

This statement exemplifies the popular belief that "natural" or "organic" products are superior to, and safer than, "chemical" products. A quick look on the Internet reveals advertisements for "chemical-free organic" fertilizers, compost, pesticides, lawns, sheep, paint, nail polish, sesame oil, diapers, and even mattresses. In every aspect of our lives, we are bombarded with the message that chemicals are bad and organic products are natural and safe.

The Reality

Before we can understand the "organic vs. chemical" controversy, we need to clarify a few terms:

- CHEMICAL General dictionaries aren't really helpful with this definition. What is important to realize is that everything on earth, natural or otherwise, is composed of chemicals.

- ORGANIC In chemistry, this refers to any chemical compound, natural or synthetic, that contains carbon.

- ORGANIC FARMING The above definition of "organic" does not apply in this context. Instead, "organic farming" is partially defined as farming using only naturally occurring, rather than synthetic, chemicals. Therefore, "chemical-free and organic" is an oxymoron, whether in chemistry or organic farming. (In a Google search, I did not find one .edu site with the phrase "chemical-free organic;" I did find 304 .com sites, however.)

- PESTICIDE Any chemical, natural or synthetic, with the ability to kill a pest organism. Herbicides, insecticides, and fungicides kill plants, insects, and fungi, respectively. The

> use of the terms "chemical-free" or "nonchemical" in reference to any pesticide is illogical. (No .edu sites contain such language (except anecdotally), but forty-five .com sites do.)

The perception of organic superiority is also common in health-food literature; "organic" or "natural" sources of sugar (like fruit juice or honey) are promoted as being healthier than refined sugar. In fact, your body's enzymes don't recognize the difference between processed and unprocessed sucrose (or fructose). Any claims about the health benefits of trace substances associated with "natural" sugars are unsubstantiated.

In much the same way, living organisms in a landscape don't distinguish between nitrate from compost or from a bag of conventional fertilizer. It's simply a usable form of nitrogen. The other components of a nutritional amendment might be beneficial or neutral or even harmful. All components of conventional fertilizers are listed on the bag; we have no such information on compost content. Whatever the nutrient source, if too much is added to a landscape, excess nutrients will leach away from the site and increase the nutrient load elsewhere. (My correspondent also wrote, "Home gardeners . . . don't farm enough land to pollute the water." Unfortunately, this just isn't true. According to the EPA, home owners use approximately ten times more chemicals per unit area of land than farmers do. In urban areas, this is obviously a major contributor to nonpoint source pollution.)

Lest I be mistaken for encouraging the indiscriminate use of conventional landscape chemicals, let me state that I avoid using any chemical in the landscape unless absolutely necessary. I fertilize my landscape plants when they show signs of nitrogen deficiency (the most common nutrient deficiency), and I use Roundup (sparingly) to reduce massive weed problems to a more manageable size.

So why do we think that "organic" is synonymous with "safe?" It's true that naturally derived, organic products have a low environmental persistence, meaning that they are quickly broken down by microbes. Nature is not benign, however; microbes, plants, and other organisms manufacture toxins, mutagens, and carcinogens as defensive strategies. To assume that products derived from biological sources can never pose a threat to human or ecosystem health is misguided and dangerous.

The Bottom Line

- It's not important whether a chemical is natural or synthetic. What is important is knowing the properties (like toxicity and environmental persistence) of chemicals we apply to landscapes.

- Any organic substance, natural or synthetic, can cause environmental problems when added in excess of what a landscape system can absorb and utilize.

- Be conservative in what chemicals you add to a landscape, regardless of their source.

References

Grava, J., and W. E. Fenster. 1979. "Fertility levels of Minnesota lawn and garden soils, 1972–76." *Minnesota Agricultural Experiment Station Report* 1979(167).

He, Z. L., A. K. Alva, P. Yan, Y. C. Li, D. V. Calvert, P. J. Stoffella, and D. J. Banks. 2000. "Nitrogen mineralization and transformation from composts and biosolids during field incubation in a sandy soil." *Soil Science* 165(2): 161–69.

Ohmura, K., and N. Sakamoto. 2000. "Runoff of nitrate nitrogen from protected flower gardening and counterplan of reduced damage to environment." *Bulletin of Hokkaido Prefectural Agricultural Experiment Stations* 2000(79): 59–66.

Pant, H. K., P. Mislevy, and J. E. Rechcigl. 2004. "Effects of phosphorus and potassium on forage nutritive value and quantity: Environmental implications." *Agronomy Journal* 96(5): 1299–305.

Stehouwer, R. C., and K. MacNeal. 2003. "Use of yard trimmings compost for restoration of saline soil incineration ash." *Compost Science and Utilization* 11(1): 51–60.

U.S. Environmental Protection Agency. "Pesticides Industry Sales and Usage: 2000 and 2001 Market Estimates." http://www.epa.gov/oppbead1/pestsales/01pestsales/market__estimates2001.pdf. Accessed March 26, 2007.

U.S. Fish and Wildlife Service. "Homeowner's Guide to Protecting Frogs—Lawn & Garden Care." http://www.fws.gov/contaminants/Documents/Homeowners__Guide__Frogs.pdf. Accessed March 26, 2007.

Zhang, Y. P., X. Y. Lin, Y. S. Zhang, S. J. Zheng, and G. D. Zhou. 2003. "Investigating on the nutrient status and plant nutrient-limiting factors of vegetable garden soils in the suburb of Hangzhou." *Journal of Zhejiang University Agriculture and Life Sciences* 29(3): 244–50.

Original article posted in November 2001.

THE MYTH OF ORGANIC SUPERIORITY, PART 2

The Myth

"Botanically derived pesticides are safer than synthetics."

Ever since the advent of synthetic pesticides in the 1930s, we have grown increasingly wary of using these substances on our landscapes. Excessive use of DDT and other persistent pesticides has left a legacy of environmental damage and created populations of pesticide-resistant pests. We have since rediscovered the wide array of natural pesticides found in the microbial and plant worlds that were recognized and used before civilization deemed them primitive. Not only can these natural alternatives be purchased, but we can also make "home brews" by following the numerous recipes available on the Web and elsewhere. If it's good enough for nature, isn't it good enough for us?

The Reality

There is a wealth of useful information on botanically derived pesticides on the Web, and it would be redundant to repeat it here. Instead, I think it's important to consider why plants make these substances in the first place and what the implications are regarding their use.

Plants manufacture an enormous variety of chemicals; this is their line of defense against predators, parasites, and competitors. Unlike most animals, plants are pretty much stuck in their environment and cannot escape suboptimal conditions except through reproduction. Instead, they use the solar energy they've harnessed to manufacture not only sugars but various protective compounds as well. "Nature red in tooth and claw" could be reworded for the plant kingdom as "Nature red in leaf and root." It's safe to say that we haven't even scratched the surface in identifying and characterizing all the plant-derived defense chemicals that exist.

The principal chemical families known to have biocidal properties (the ability to kill living organisms) are the alkaloids (including nicotine, ryania, and sabadilla); the terpenoids (including neem and pyrethrins); and the flavonoids (including rotenone). These commonly used and easily available compounds are discussed briefly below.

- Nicotine is a well-known alkaloid extracted from the leaves of *Nicotiana* and once easily available as nicotine sulfate. This highly neurotoxic chemical is generally not available for home use and poses a threat to any animal that inhales or touches it. Unfortunately, there are still publications and Web sites encouraging the use of "tobacco teas," but brewing these decoctions should be avoided. Once in the soil environment, this nitrogen-rich compound is quickly broken down by microbes in a matter of hours.

- Ryania is a mixture of compounds extracted from the roots and stems of the tropical plant *Ryania speciosa*. The principle active ingredients in ryania are ryanodine and related alkaloids. It is a relatively selective, ingestible, neurotoxic insecticide with low-to-moderate toxicity for birds, fish, and mammals. It is fairly persistent in the environment, though its environmental biodegradation is not yet well understood.

- Sabadilla is extracted from the seeds of the tropical genus *Schoenocaulon*. Two alkaloids comprise the active ingredients of this extract, which, like other alkaloids, have a neurotoxic effect upon insects. Sabadilla works either as an ingestible or contact insecticide and unfortunately affects bees as well as targeted pests. It has very low toxicity to mammals and is not persistent in the environment.

- Neem is a mixture of chemicals extracted from the seeds (and sometimes the leaves and bark) of the Asian tree *Azadirachta indica*. The principle active ingredient of neem is the bitter terpenoid azadirachtin, which is an insect-feeding deterrent and a growth regulator. Neem extracts prevent insects from maturing and completing their lifecycle, reducing the local insect population. It has very low toxicity to mammals and is not persistent in the environment.

- Pyrethrins belong to the terpenoid family and were originally extracted from chrysanthemum flowers. Once ingested, these compounds affect the nervous systems of a broad range of insects, incapacitating or killing them.

Though short-lived, natural pyrethrins are extremely toxic to fish and bees and somewhat toxic to birds, but pose little hazard to mammals. Some insects have the ability to detoxify pyrethrins and are therefore resistant to the natural compounds. Synthetic pyrethroids have been developed that are more effective against insects and less toxic to other life forms.

- Rotenone is a flavonoid (commonly and erroneously identified as an alkaloid) extracted from the roots of a number of different tropical legumes, including *Derris* and *Lonchocarpus* species. Though it degrades quickly, it is both a contact and ingestible insecticide that kills a wide range of insects, including beneficials. It is highly toxic to fish and slightly toxic to waterfowl. Furthermore, recent studies have linked chronic rotenone exposure to Parkinson's disease in humans.

All pesticides, natural or synthetic, undergo extensive testing to determine toxicities to laboratory organisms and to predict threats to ecosystems. Before any pesticide can be licensed for use, an LD50 must be established. The LD50 is the amount of the chemical necessary to cause death (the "lethal dose") in 50 percent of the test population (typically rats). Therefore, a low LD50 translates to a higher potential risk for humans and other organisms. Below are ranges of LD50s for several common organic and two synthetic pesticides, reported in milligrams of pesticide per kilogram of animal weight. These numbers vary depending on species tested. (To put this in human-risk perspective, compare the LD50 for aspirin (1,200) and for glyphosate, the active ingredient in Roundup (5,600).)

nicotine	55
Sevin (synthetic)	246–283
rotenone	132–1,500
pyrethrin	200–2,600
ryania	750–1,200
malathion (synthetic)	1,000–10,000
sabadilla	4,000–5,000
neem	>5,000

From this table alone it should be evident that botanically derived pesticides are not always safer than synthetics, and in some cases are much worse. Botanical insecticides can harm non-target species such as beneficial insects, fish, birds, and mammals. This is plant warfare, and no distinctions are made between friends and enemies.

Improper and continual use of pesticides, whether naturally or synthetically derived, will increase the likelihood that resistant pest populations will evolve. Nature is not static; to survive, organisms must constantly adapt to a changing environment, and this includes chemical exposure. The faster a species can reproduce, the more likely it is that chemically resistant populations will arise.

Instead of being so quick to use chemical controls of any sort, we should be willing to adopt the philosophies of Plant Health Care and Integrated Plant Management. By maintaining a healthy, diverse soil and plant environment, and by utilizing cultural, physical, and biological forms of pest control, we can dramatically reduce our dependence on chemicals, natural or synthetic, that by their nature will kill other organisms and weaken the stability of a landscape system.

The Bottom Line

- Botanically derived pesticides are not always "safe," and some are more hazardous than synthetics.
- Any improperly used pesticide will contaminate nearby terrestrial and aquatic systems.
- Use of broad-spectrum pesticides will kill beneficial insects, leaving plants open to attack from pests.
- Continual use of any pesticide will eventually induce pesticide resistance in pest species.

References

Robinson, T. 1991. *The Organic Constituents of Higher Plants,* 6th ed. North Amherst, MA: Cordus Press.

U.S. Environmental Protection Agency. "Azadirachtin (121701) and clarified hydrophobic extract of neem oil (025007) fact sheet." http://www.epa.gov/oppbppd1/biopesticides/ingredients/factsheets/factsheet__025007.htm. Accessed March 26, 2007.

U.S. Environmental Protection Agency. "Pyrethrins." http://www.epa.gov/REDs/pyrethrins__red.pdf. Accessed March 26, 2007.

U.S. Environmental Protection Agency. "Rotenone." http://www.epa.gov/oppsrrd1/reregistration/rotenone/. Accessed March 26, 2007.

U.S. Environmental Protection Agency. "Ryanodine." http://www.epa.gov/REDs/factsheets/2595fact.pdf. Accessed March 26, 2007.

U.S. Environmental Protection Agency. "Sabadilla alkaloids." http://www.epa.gov/oppsrrd1/reregistration/sabadilla/. Accessed March 26, 2007.

Original article posted in October 2002.

UNDERSTANDING HOW PLANTS WORK

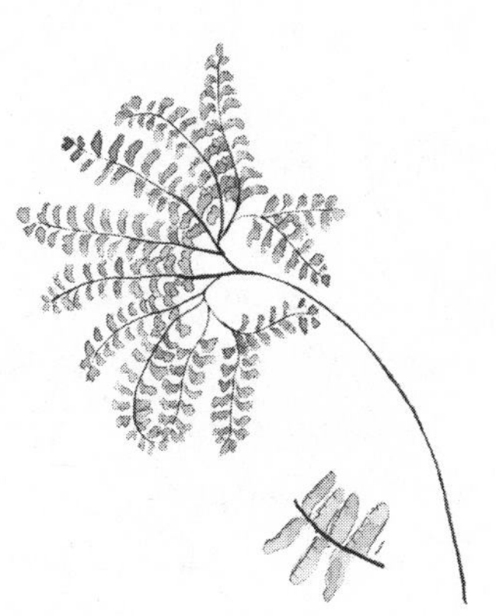

THE MYTH OF FRAGILE ROOTS

The Myth

"You shouldn't disturb the root ball when transplanting trees and shrubs."

"When you transplant, try not to disturb the roots, just take the whole pot-shaped lump of soil/roots and pop it into its new home." This and similar advice can be found on Web sites and in gardening books, all which warn us of the fragile nature of roots. When you upend a container and slide out the root ball, it's an innate response to handle those tiny white and brown strands gingerly so as not to break them. Since the survival of a newly installed tree or shrub is dependent upon healthy, functioning roots, it seems obvious that the more intact the root system, the better the chances of establishment. Anything that damages this intricate web would seem to add to transplant shock.

The Reality

Though gentle handling of roots is good advice when transplanting seedlings (especially annual flowers and vegetables), woody perennials, shrubs, and trees all benefit from a more vigorous approach. Surprisingly some of the harshest techniques result in the healthiest plants. There are several reasons for this.

A more aggressive approach to root preparation at transplant time can uncover potentially fatal root flaws. Containerized materials, especially those in gallon-sized pots, often have serious root problems as a result of poor potting-up techniques. Pot-bound plants develop circling roots systems when roots encounter the edge of the container and continue to grow along this surface. If not corrected during potting up or transplanting, these roots become woody and inflexible. Woody circling roots continue to expand in girth and can eventually girdle the trunk, leading to the early demise of otherwise healthy trees and shrubs. Circling roots and otherwise misshapen roots can be removed by careful pruning, thus eliminating future root problems before they threaten the survival of a tree or shrub. It's important to realize that roots respond to pruning in much the same way as the crown: Pruning induces new growth. Roots that are pruned at transplant time, especially those that are excessively long or misshapen, will respond by generating new, flexible roots that help establish the plant in the landscape. It is vital that new transplants are kept well watered during this time.

Another problem with containerized materials can also be avoided during your root inspection. In general, the media in the container is a soilless mix with a large proportion of organic matter and pumice. If transplanted with the plant as part of the root ball, this material will inhibit root development outside the planting hole. Roots grow best where oxygen, nutrients, and water are most plentiful, and these resources are initially greater in the porous, organic potting mix than in the native soil. But the space is limited, and these resources are quickly depleted as roots criss-cross throughout the planting hole; at this point, it is difficult, if not impossible, for roots to establish in the native soil. Furthermore, the porous texture of the planting media loses water to the environment by evaporation more rapidly than the surrounding native soil, resulting in increased water stress to your new transplant, especially in summer months. For root establishment, it is much better to bare-root the plant by removing as much of the container material as possible before the plant is installed. The best use for the discarded container mix is as a topdressing over the disturbed soil. When covered with wood chips or another mulch that will reduce weed colonization, the container media serves as a nice source of slow-release nutrients.

The Bottom Line

- Containerized plants are notorious for concealing fatal root flaws.
- Plants with woody roots often need corrective root pruning before transplanting.
- Bare-rooting container plants is a more successful transplanting technique, as root flaws can be corrected and container media removed.
- With a healthy, well-watered plant, root pruning at transplant time will induce vigorous new root growth and assist in the plant's establishment.

References

Arnold, M. A. 1996. "Mechanical correction and chemical avoidance of circling roots differentially affect post-transplant root regeneration and field establishment of container-grown shumard oak." *Journal of the American Society for Horticultural Science* 121(2): 258–63.

Harris, R. W. 1968. "Factors influencing root development of container-grown trees." *Proceedings of the 1967 International Shade Tree Conference*, 43:304–14.

Harris, R. W., and W. B. Davis. 1971. "Effects of root pruning and time of transplanting in nursery liner production." *California Agriculture* 25(12): 8–10.

Keever, G. J., and G. S. Cobb. 1988. "Optimizing production of container-grown pecans." *Alabama Agricultural Experiment Station Research Report Series* 5:10–11, 16.

Watson, G. W., and S. Clark. 1993. "Regeneration of girdling roots after removal." *Journal of Arboriculture* 19(5): 278–80.

Original article posted in January 2003.

PRUNING FLAWED WOODY ROOTS BEFORE TRANSPLANTING

- Prepare a work space in a cool, shaded environment; have water available at all times.
- Remove all foreign materials from roots as described in "The Myth of Collapsing Root Balls."
- From this point on, keep roots submerged in water at every step except when actually pruning.
- Ignore fibrous roots for now; focus on woody roots that can't be straightened with your hands.
- Your root system should extend radially and horizontally like spokes on a bicycle wheel. You will be removing woody roots that do not follow this appearance.
- Investigate the very center of the root mass—this is the area that represents the earliest root system and is often poorly structured. Remove any large masses of fused root tissue that do not terminate in normal woody roots.
- After you have cleaned up the center of the root mass, look to the areas where roots may have encountered a container edge and made at least a ninety-degree turn inward or downward. Remove the roots at this point.

Continued on next page

- The woody roots should now form a horizontal and radial structure. Shorten any excessively long, fibrous roots that will be difficult to transplant.
- Your plant is now ready to transplant following the instructions in "The Myth of Collapsing Root Balls."
- Be aware that the more root material you remove, the longer it will take your plant to recover and establish. I have successfully established good-sized trees from which I've removed up to 75 percent of the root system. During this time, it is critical to keep the root zone adequately hydrated and to provide shading of the crown, if possible, if the plant is in direct sunlight.
- Patience is key now; existing leaves will probably die, and new leaves will be slow to appear. I have seen some heavily root-pruned plants take several months to initiate leaf growth. Give the plant a chance to develop its new root system; when it has established, you will be rewarded with a flush of new leaves.

THE MYTH OF MIGHTY ROOTS

The Myth

"Wire baskets will not interfere with the root growth of transplanted trees."

What to do with those wire baskets that so neatly restrain the root balls of balled and burlapped trees? A perusal of Web sites reveals that many nurseries and plant advice pages recommend leaving the wire basket intact or loosening it or bending the top portion down so it's covered by soil. This "out of sight, out of mind" approach can give customers a false sense of security; when their tree begins to fail a few years down the road, most likely they won't associate the failure with potential root problems. Why is there such a reluctance to remove this slow-to-degrade, impervious material from planting holes?

About ten to fifteen years ago, a handful of papers were published on this topic by Dr. Glen Lumis at the University of Guelph. His most-often-cited work, published in 1990 in the popular magazine *American Nurseryman*, focused on *Salix* and *Acer* species planted in wire baskets in a municipal park and exhumed four to fifteen years later. Lumis reported that the roots of these trees were able to grow around and over the wire with no permanent girdling. This single paper (which neither contains much scientific information nor explains why these trees were removed in the first place) has been used to justify the practice of leaving wire baskets on transplanted trees.

The Reality

There are at least five publications that conclude that wire baskets do interfere with root growth or cause girdling or other root problems. In fact, one of these publications was coauthored by Lumis in 1992 in the *Journal of Arboriculture*. In contrast to the findings in his earlier paper, Lumis's later paper reported that only *Populus* whips were able to overcome wire girdling; *Celtis* and *Fraxinus* roots were compromised. It appears that this may be a species- and site-specific issue. *Populus* and *Salix* have aggressive roots that can grow through anything, so it's not surprising that wire baskets present no challenge to these genera. Other trees are apparently much more sensitive to girdling by wires, and it stands to reason that poorer site conditions will translate into increased stress on the tree. In other words, a willow growing in a large park setting will have fewer environmental stresses than an ash grown as an urban street tree and would therefore be less likely to be hampered by underground impediments.

This and the other four papers in peer-reviewed journals are ignored by the Web sites and businesses who instead champion the earlier article in *American Nurseryman*. I can only assume that the reason for this is to make transplanting appear faster and easier and therefore more desirable to consumers.

The Bottom Line

- Most long-term research has demonstrated that wire baskets are harmful to tree root establishment.
- Wire baskets should be removed from root balls prior to planting trees.

References

Appleton, B., and S. A. Floyd. 2004. "Wire baskets—current products and their handling at planting." *Journal of Arboriculture* 30(4): 261–65.

Feucht, J. R. 1993. "Update on utilizing wire baskets in the green industry." *Long Island Horticulture News* (February): 2–3.

Feucht, J. R. 1986. "Wire baskets can be slow killers of trees." *American Nurseryman* 163(6): 156–59.

Goodwin, C., and G. Lumis. 1992. "Embedded wire in tree roots: Implications for tree growth and root function." *Journal of Arboriculture* 18(3): 115–23.

Kuhns, M. R. 1997. "Penetration of treated and untreated burlap by roots of balled-and-burlapped Norway maples." *Journal of Arboriculture* 23(1): 1–7.

Lumis, G. P. 1990. "Wire baskets: a further look." *American Nurseryman* 172(5): 128–31.

Lumis, G. P., and S. A. Struger. 1988. "Root tissue development around wire-basket transplant containers." *HortScience* 23(2): 401.

Original article posted in December 2001.

THE MYTH OF TOP-PRUNING TRANSPLANTED MATERIAL

The Myth

"Transplanted plants should have their crowns pruned to compensate for reduced root mass."

Gardeners are often advised to prune back the crown of transplanted trees and shrubs by as much as 50 percent to reduce transpiration and compensate for lost root systems. Internet Web sites, even the usually reliable .edu sites, continue to spread this myth. This is a common practice in nurseries, where top-pruning of containerized plants reduces shipping costs and has been shown to increase tree survival and growth under nursery conditions. Research performed under landscape conditions, however, shows that pruning when transplanting is not necessarily beneficial and may even harm the tree.

For example, one study demonstrated that water usage actually increased in several top-pruned tree species after the first five to six weeks in response to new shoot growth. Other studies have reported stunted growth or poor establishment as a result of top-pruning. But conflicting evidence exists regarding the costs and benefits of crown pruning. It is important to keep in mind that there are short-term and long-term effects of top-pruning. While there might be immediate benefits in terms of balancing root and shoot mass, the long-term effects are generally negative. This difference in timescale probably accounts for much of the contradictory evidence in the scientific literature; many studies are relatively short-term.

The Reality

When growing tips are removed from most plants, the immediate response is bud break (or leafing out) below the cuts. This results in a bushier plant and can destroy the natural form of young decurrent (rounded) and excurrent (pyramidal) trees. Such growth requires that the plant put its resources into top growth at the expense of root growth. Gardeners see the top part of the plant growing, but are unaware that the root growth has been inhibited by the reallocation of energy resources. Plants left unpruned after transplanting appear to be dormant to the untrained eye—but they actually are putting resources into root growth. When the roots have become established, shoot growth resumes.

Obviously, the practice of pruning the crown of a transplanted tree or shrub will cause the plant to respond in a way that is neither desirable nor healthy. In addition to interfering with the plant's ability to establish its roots, the removal of a significant portion of the crown also means the plant has lost biomass and cannot photosynthesize at its previous level. Thus, plants that have been top-pruned are hit with a double whammy: part of their photosynthetic system is removed, and those resources that are left are directed toward new shoot development. It's no surprise that root establishment under these conditions is difficult.

The Bottom Line

- There is no need to top-prune landscape plants if post-transplant irrigation is available (and all new landscapes need post-transplant irrigation!).

- The only time transplanted materials should be pruned is to remove broken, dead, or diseased branches, or to make structural corrections to young trees.

- If pruning is warranted, use thinning rather than heading cuts to preserve tree structure.

References

Bayala, J., Z. Teklehaimanot, and S. J. Ouedraogo. 2004. "Fine root distribution of pruned trees and associated crops in a parkland system in Burkina Faso." *Agroforestry Systems* 60(1): 13–26.

Jones, M., and F. L. Sinclair. 1996. "Differences in root system responses of two semi-arid tree species to crown pruning." *Agroforestry Forum* 7(2): 24–27.

Jones, M., F. L. Sinclair, and V. L. Grime. 1998. "Effect of tree spe-

cies and crown pruning on root length and soil water content in semi-arid agroforestry." *Plant and Soil* 201(2): 197–207.
Watson, G. W. 1998. "Tree growth after trenching and compensatory crown pruning." *Journal of Arboriculture* 24(1): 47–53.

Original article posted in October 2000.

THE MYTH OF TREE TOPPING

The Myth

"It's like a haircut—sometimes it's necessary, and a tree can always grow out of a bad one."

Since plant scientists and arborists unanimously agree that tree topping is an unjustifiable tree management practice, you might assume that the word would have trickled down to practitioners and their customers. Yet every year brings a new crop of buzz-cut trees. It also brings a new crop of excuses (culled from the Internet):

- "I want to trim the top branches off a 75' tall maple because it's causing excess shade in my yard. I want the tree to live, but just be smaller."

- "I wouldn't make the sweeping generalization that all tree topping is bad. . . . Locals here whack their weeping willows every few years and those trees seem to relish the opportunity to fill out again."

- "It is necessary for the electric company to top trees that grow into the power lines."

- "The trees look like hell for a while but seem to get used to the treatment."

A tree service company states, "Although topping a tree is not usually recommended, it is sometimes very necessary. Some of the time it can be a definite safety issue. Other times a tree is topped to get rid of mistletoe." There's a Web page entitled "Trees that love chainsaws"! And, in a questionable marketing move, a UK company has trademarked the name "tree-topping" to describe its approach to forest management: After thinning, the remaining trees are topped to "reduce wind throw."

The Reality

I'll preface this discussion with a caution that I am referring to pruning trees (not shrubs or hedges), and only trees that are being maintained in their natural form. There are many types of formal pruning techniques, including pollarding, pleaching, espaliering, et cetera, but they are not included in this discussion.

A reduction cut (also called "thinning to a lateral") is a method of pruning used to reduce the height of a tree. When done properly, branches are cut back to a lateral branch at least one-third the diameter of the limb being removed. The lateral branch becomes the source of new terminal growth, and subsequently the tree maintains a natural form. This is an appropriate pruning technique for decurrent (rounded) trees, but should never be used on excurrent (pyramidal) trees except to remove multiple leaders.

Unfortunately, many tree cutters (certainly not certified arborists!) claim to thin to laterals when in reality they are topping the tree. Also known as "hatracking," "height reduction," "canopy reduction," "heading back," or "stubbing back," this type of pruning cut removes a terminal shoot back to a point where there is no appropriate lateral branch to take over the terminal role. In response, multiple shoots (or leaders) begin to compete for dominance, resulting in the infamous "hydra" look. What has now been created is a high-maintenance, potentially hazardous tree that must be constantly pruned. Pruning a tree yearly is certainly not environmentally sustainable or cost-effective—but it does keep tree cutters in business!

There are plant health issues with tree topping; it's been demonstrated that sun damage, nutrient stress, insect attack, and decay result from unnecessary and incorrect pruning procedures. There are also aesthetic issues with tree topping; improperly pruned trees are ugly. For years, groups such as the International Society of Arboriculture and the Seattle-based PlantAmnesty have tried to educate professionals and home owners about the horrors of tree topping, from both a plant health and an aesthetic perspective; yet tree topping continues. Perhaps what's needed in today's tort-happy society is a liability perspective to make tree cutters and those that hire them sit up and take notice.

After topping, many epicormic shoots (from dormant buds beneath the trunk surface) arise and develop into weakly attached branches. These branches, many of which can become leaders, continue to develop girth and weight and have an increasing potential to fall and cause damage to people or property. From a legal standpoint, the owner of such a tree is responsible for damages if it can be proved that the owner was negligent. If I were to tell my neighbor that her tree constituted a hazard, and later this same tree fell and damaged my property, in some states I would be entitled to both actual and punitive damages. There is no doubt within scientific and arborist communities that incorrect pruning can cause trees to become hazardous. Only one expert witness is needed to demonstrate this, and the owner, or the landscape maintenance company, will be found responsible.

If every property owner was given this last paragraph of information, I would bet that tree topping would come to a screeching halt. But as long as anyone with a pickup truck and a chainsaw is allowed to call himself a "landscape professional," property owners by and large will remain blissfully unaware. Property owners need to become educated: They need to insist on certified arborists for tree care and make wise decisions before installing plant material that will outgrow its welcome.

The Bottom Line

- Tree topping is never a justifiable pruning practice; it increases tree health problems and is aesthetically unappealing.

- Certified arborists and other legitimate landscape professionals do not practice tree topping.

- There are acceptable pruning techniques designed to keep trees away from power lines and other structures.

- A topped tree will require constant maintenance and has an increased potential to become hazardous.

- Hazardous trees are a liability, and ultimately the property owner is responsible for any damage hazard trees cause.

- If problems caused by a tree cannot be solved through acceptable management practices, the tree should be removed and replaced with plant material more appropriate for the site.

- Think about the mature size of a tree and where it will grow relative to power lines and other structures before you plant it.

References

Close, D. D., J. W. Groninger, J. C. Mangun, and P. L. Roth. 2001. "Home owners' opinions on the practice and effects of topping trees." *Journal of Arboriculture* 27(3): 160–65.

Fahad, R., H. Shahid, and I. Q. Bhabha. 2005. "Phyto-sociological study and determination of carrying capacity of the Reserve Forest compartment-17 of Margallah Hills National Park." *Pakistan Journal of Agricultural Sciences* 42(1–2): 71–74.

Fazio, J. R., and E. E. Krumpe. 1999. "Underlying beliefs and attitudes about topping trees." *Journal of Arboriculture* 25(4): 193–98.

Gilman, E. F., and G. W. Knox. 2005. "Pruning type affects decay and structure of crapemyrtle." *Journal of Arboriculture* 31(1): 48–53.

Ho, R. H., and H. O. Schooley. 1995. "A review of tree crown management in conifer orchards." *The Forestry Chronicle* 71(3): 310–16.

Kaiser, C. A., M. L. Witt, J. R. Hartman, R. E. McNiel, and W. C. Dunwell. 1986. "Warning: Topping is hazardous to your tree's health!" *Journal of Arboriculture* 12(2): 50–52.

Karlovich, D. A., J. W. Groninger, and D. D. Close. 2000. "Tree condition associated with topping in southern Illinois communities." *Journal of Arboriculture* 26(2): 87–91.

Stamen, R. S. 1994. "Are you a candidate for a lawsuit?" *Arbor Age* (July 1994): 10–12.

Toussaint, A., V. K. de Meerendre, B. Delcroix, and J. P. Baudoin. 2002. "Analysis of physiological and economical impacts of the roadside trees pruning." *Biotechnologie, Agronomie, Societe et Environnement* 6(2): 99–107.

Original article posted in September 2003.

HOW BIG IS BIG?

How often have you planted a tree or shrub only to have it outgrow its space? Overgrown trees and shrubs obstruct views, traffic, and pedestrians, turning what should be a landscape enhancement into a nuisance. It's especially frustrating when you've carefully read nursery tags to avoid this very problem. Why is it so hard to predict how big trees and shrubs will grow?

There are at least four interacting factors that will influence the mature size of any particular plant species:

- Genetics plays a part in determining final plant size; there may be cultivars bred for smaller sizes more appropriate for urban landscapes.
- Regional climate plays a major role in controlling plant growth; the more limiting the environmental conditions, the smaller the mature size, regardless of genetic predisposition.
- Geographic features within a climate region will further affect plant size; there will be differences in the heights of plants found on the north-facing slope of a hill compared to those on the south-facing slope.
- Site microclimate, imposed by differences in soils, or by structures that influence water movement, temperature, light, and wind, will have an impact on mature size.

The caveat here is not to believe what nursery tags report as mature size. Instead, observe the same species in your location to get an idea of the range of heights possible. You can also surf the Web to find out how your species of interest performs in similar climates around the world. When you do select plant species, anticipate how they will change—and change the landscape—over many years.

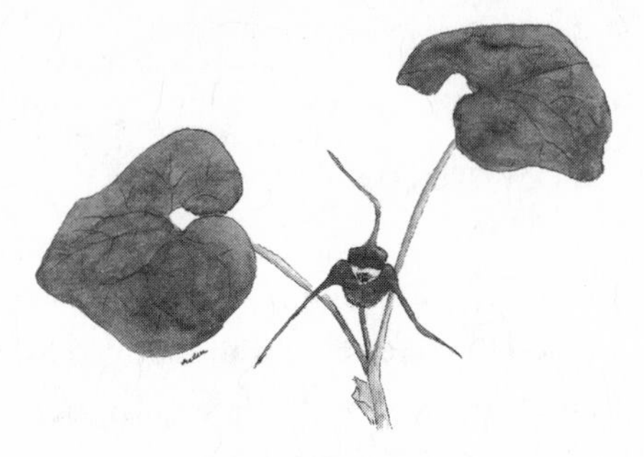

THE MYTH OF HOT-WEATHER WATERING

The Myth

"Watering plants on a hot sunny day will scorch their leaves."

Every year, as we enter the hot summer months, another bit of well-meaning advice rears its head. Magazine articles, books, and Web sites all warn against watering plants during the heat of the day. They claim that the water drops accumulating on leaf surfaces act as tiny magnifying glasses, focusing the sun's energy into intense beams that burn the leaf. Furthermore, we're told that since water efficiently conducts heat, wet leaves are more likely to burn than dry ones. This all sounds very plausible (it has the patina of physics, after all), and there is anecdotal evidence that seems to support a cause-and-effect relationship between midday watering and leaf dieback.

The Reality

This is one of those myths that refuses to die. Although most (but not all!) of the .edu Web sites I checked dispel this myth, hundreds of other domains on the Web keep the misinformation alive. If your plants are showing signs of water stress in the middle of the day, by all means you should water them! Postponing irrigation until the evening (not a good time to water anyway, as this can encourage fungal pathogens) or the following morning could damage your plants and expose them to opportunistic diseases.

There are many causes of leaf scorch, but irrigation with fresh water is certainly not one of them. Hundreds of scientific publications on crop plants, turf, woody shrubs, and trees have examined foliar scorch, and not one of them has implicated midday irrigation as a causal agent. What does cause damage, however, is too little or too much water, either of which can result in tip and marginal leaf scorch, shoot dieback, stunted growth, and leaf abscission. Drought is the most common cause of leaf scorch. After drought, the most common source of these problems is salt, in particular salts containing sodium (Na^+) and/or chloride (Cl^-).

Salt can enter a plant's microhabitat via spray from the ocean or other salt water bodies, or by runoff from road de-icing salts. More widespread damage occurs with salts in soils or irrigation water. Some of these salts come from the overapplication of fertilizers, herbicides, and insecticides. Others are naturally found in irrigation water that runs through particularly saline soils. Salts in the soil tend to be a more significant problem in arid climates. In urban areas, irrigation with recycled or gray water can add toxic levels of salt to a plant's environment.

Plants that are not adapted to dry or saline environments have a difficult time extracting pure water from a salty solution. We all know the dangers of drinking seawater, and we know what happens when we put salt on slugs in the garden. The dehydration and death that you would expect in these examples also occur in plant tissues, and in particular those tissues that transpire the most water—the leaves. Salts can either injure roots directly, reducing their ability to function and thus reducing water transport to the leaves, or they can accumulate in the leaves themselves. In either case, water loss occurs at the tips and margins of the leaves and, if not treated promptly, will lead to tip and marginal necrosis.

Besides drought and salt, other causes of leaf scorch include wind stress, high temperatures, reflected light, and cold stress. All of these environmental stressors are directly linked to decreasing water availability in leaves. Poor root health, imperiled by soil compaction, flooding, or restricted space, will also induce leaf scorch. Lack of foliar potassium (the "K" in fertilizers) prevents leaves from regulating stomatal openings and leads to higher water loss. Urea, contained in some fertilizers and in urine, can burn foliage and is a common cause of turf damage. Regardless of the cause, leaves deficient in water have been shown to be more susceptible to opportunistic pests and pathogens, including mites and fungal leaf scorch.

To prevent leaf scorch, it's important to have environmental conditions conducive to optimal root health—adequate moisture, oxygen, space, temperature, and nutrients are part of a healthy root zone. Some studies have found that additional nitrogen helps prevent leaf scorch (perhaps by increasing root growth and uptake capabilities). It's also crucial to watch foliage for signs of water stress, generally seen first in the wilting of young stems and leaves. Once leaf tissues have passed the terminal wilt stage, no amount of water will save them. People who don't recognize the signs of terminal wilt and add water anyway might then associate their midday watering with the marginal and tip leaf burn that follows. Consider the plant's needs in terms of sun/shade requirements; a shade-loving plant in an area with high light exposure, reflected heat, wind, or temperature extremes is going to show leaf burn on a continuing basis.

The Bottom Line

- Wet foliage is not susceptible to sunburn.
- Any time plants exhibit drought stress symptoms is the time to water them.
- Optimal watering time is in the early morning; watering during the day increases evaporative loss, and evening watering can encourage the establishment of fungal pathogens.
- Do not overuse fertilizers and pesticides, especially those containing sodium or chloride salts.
- If using recycled or gray water, consider running the water through a filtering system before applying it to plants.
- Analyze site conditions to ensure optimal root and shoot health and to prevent drought problems.

References

Burlo-Carbonell, F., A. Carbonell-Barrachina, A. Vidal-Roig, and J. Mataix-Beneyto. 1997. "Effects of irrigation water quality on loquat plant nutrition: Sensitivity of loquat plant to salinity." *Journal of Plant Nutrition* 20(1): 119–30.

Cregg, B. M., and M. E. Dix. 2001. "Tree moisture stress and insect damage in urban areas in relation to heat island effects." *Journal of Arboriculture* 27(1): 8–17.

McElrone, A. J., J. L. Sherald, and I. N. Forseth. 2001. "Effects of water stress on symptomatology and growth of *Parthenocissus quinquefolia* infected by *Xylella fastidiosa*." *Plant Disease* 85(11): 1160–64.

McNab, S. C., D. G. Williams, and P. H. Jerie. 1994. "Effect of intensity and duration of twospotted spider mite (Acari: Tetranychidae) infestation and water stress on leaf scorch damage of 'Bartlett' pear." *Journal of Economic Entomology* 87(6): 1608–15.

Nielson, R. F., and O. S. Cannon. 1975. "Sprinkling [crops] with salty well water can cause problems [from foliar burn, injuries]." *Utah Science* 36(2): 61–63.

Niu, G. H., and D. S. Rodriguez. 2006. "Relative salt tolerance of selected herbaceous perennials and groundcovers." *Scientia Horticulturae* 110(4): 352–58.

Opit, G. P., G. K. Fitch, D. C. Margolies, J. R. Nechols, and K. A. Williams. 2006. "Overhead and drip-tube irrigation affect twospotted spider mites and their biological control by a predatory mite on impatiens." *HortScience* 41(3): 691–94.

Riccioni, L. 2004. "Chrysanthemum white rust, quarantine pathology." *Colture Protette* 33(5): 79–81.

Romero-Aranda, R., and J. P. Syvertsen. 1996. "The influence of foliar-applied urea nitrogen and saline solutions on net gas exchange of citrus leaves." *Journal of the American Society for Horticultural Science* 121(3): 501–6.

Samra, J. S. 1989. "Effect of irrigation water and soil sodicity on the performance and leaf nutrient composition of mango cultivars." *Acta Horticulturae* 231:306–11.

Wood, B. W., C. C. Reilly, and W. L. Tedders. 1995. "Relative susceptibility of pecan cultivars to fungal leaf scorch and its interaction with irrigation." *HortScience* 30(1): 83–85.

Original article posted in August 2002.

HOW/WHAT/WHEN/ WHERE TO PLANT

THE MYTH OF INSTANT LANDSCAPING

The Myth

"How hard can it be to stick a plant in the ground?"

Recently I stumbled across a landscaping show on public television. The host, an enthusiastic and personable landscape designer, performed a landscape makeover on a neglected yard. The design and plant selections were fine, but the trouble began when he demonstrated his installation technique. A hole was dug to the same size as the container, the plant was removed and inserted directly into the hole, and soil was mounded up to the base of the trunk. It took all of fifteen seconds, giving viewers the impression that plant installation is a snap and that a beautiful landscape would result.

The Reality

It takes more than fifteen seconds to properly install a containerized plant. Since unobstructed roots grow horizontally, it's important to direct them outward upon installation. Pot-bound plants possess circling roots which, unless straightened, will continue to circle, decreasing plant stability and increasing the likelihood that girdling and death will eventually occur. For this reason, the planting hole should be at least twice as wide as the container to allow proper root spread. A small mound of soil at the bottom of the planting hole will support the root crown and allow the plant to remain at grade (or slightly higher).

It seems like common sense to add the rich, well-drained potting material to the planting hole to give the plant a head start. In actuality, this contributes to one of the leading causes of post-installation plant death. The potting material is always more porous than the surrounding soil; hence, it dries out faster and needlessly stresses the roots of the plant. The next time you are inspecting a newly installed, suffering plant, stick your finger in the planting hole. It's probably pretty dry. Furthermore, even well-watered plants in these situations are slower to establish roots outside the planting hole because of the textural barrier created between the potting material and the soil.

At the University of Washington, my students and I investigated the effects of removing container media from roots prior to installing trees and shrubs into the landscape. We found no negative impact from bare-rooting these plants, and I now recommend this practice for all trees and shrubs. In bare-rooting, potting material is shaken off the root system; in the case of pot-bound plants, roots may need to be loosened in a bucket of water. It is critical to keep the roots moist from this point onward, as it is now a bare-root plant. An added bonus of removing the potting material is that root defects can be identified and corrected prior to installation. After proper positioning of the plant, soil is backfilled, water is added, and a thick topdressing of mulch completes the installation.

Installing a plant following these guidelines takes longer, especially if the root system is a tangled woody mess. The long-term benefit, however, is a healthier plant that establishes quickly in the landscape and will require less aftercare than one that is simply popped and dropped.

The Bottom Line

- A planting hole should be at least twice as wide but no deeper than the root mass.
- Pot-bound plants need to have circling roots straightened or removed.
- Container potting material needs to be removed from the root mass and composted—not added to the planting hole.
- Roots need to be teased apart and directed outward.
- Plants that are installed correctly will require less aftercare and have a longer life in the landscape.

References

Chalker-Scott, L. 2005. "Growing healthier trees." *Organic Gardening* (Oct./Nov.): 14–15.

Chalker-Scott, L. 2005. "Transplant science: Will the patient live?" *American Nurseryman* 201(6): 20–23.

Harris, R. W., J. R. Clark, and N. P Matheny. 2004. *Arboriculture*, 4th ed. Upper Saddle River, NJ: Prentice Hall.

Stout, T. 2004. "Treating containerized material as bare-root material: A comparison of installation techniques." Master's thesis, University of Washington.

Original article posted in August 2001.

THE MYTH OF NATIVE PLANT SUPERIORITY

The Myth

"Always choose native plants for environmentally sustainable landscaping."

As natural ecosystems continue to shrink, people have become more interested in native plants and landscapes. This admirable dedication to our natural heritage has manifested itself in native gardens springing up in every place imaginable. Yet at the same time, I see more of these native gardens suffering from disease, pests, and general decline. What's happening? Aren't native plants supposed to be resistant to local pathogens and parasites?

The Reality

There are some urban areas where many native plants just can not survive (or do so only with substantial maintenance). Such areas can include parking strips, traffic circles, and parking lots—in short, areas with limited soil area and lots of environmental stress. Consider the realities of these landscapes:

- discontinuous, dissimilar layers of topsoils and subsoils with poor drainage and aeration
- significant compaction and other physical disturbances as a result of animal, pedestrian, and vehicular traffic
- alkaline pH due to leaching of lime from concrete
- inadequate or improper fertilizer application
- lack of mulch or other soil protection
- lack of adequate water in summer months
- increased heat load from asphalt reflectance
- air pollution

Many of the trees and shrubs native to the Northwest region evolved in thin, acidic soils with adequate moisture. When these species are installed in urban landscapes with significantly different soil and water characteristics, they are challenged by a new set of environmental circumstances. As landscape plantings begin to suffer from multiple stresses, they become prone to invasion from opportunistic insects, bacteria, and fungi. Stress can weaken a plant's natural resistance to local pests: witness the recent decline in the Northwest's native *Arbutus menziesii* (Pacific madrone) populations.

Another example of the failure of native trees to survive in urban sites comes from Palm Desert, California. Many of the parking lots in Palm Desert were planted with native mesquite. Mesquite survives in its arid environment by developing both a deep taproot and an extensive shallow root system. When planted into the very limited soil spaces typical of parking-lot tree wells, these trees often tilt or topple as a result of insufficient lateral root development. The City of Palm Desert has recently looked to nonnative tree species, including ash, to replace mesquite in these settings.

The Bottom Line

- Site considerations should always dictate plant selection.

- Native, temperate forest plants are excellent choices for temperate, unrestricted sites with acidic, well-drained soils and adequate precipitation (i.e., greater than thirty inches a year).

- For sites with limited, alkaline, and/or poorly drained soils, choose species adapted to environments with similar soils. Consider especially those species that tolerate clay soils.

- For sites exposed to increased heat load, choose species adapted to hot, dry climates that can also tolerate cool, wet winters.

- Instead of installing large trees in limited sites, consider smaller trees or shrubs that can be arborized.

- Be sure to protect soils with mulch, especially where foot traffic causes compaction.

References

Adams, A. B., and C. W. Hamilton, eds. 1999. *The Decline of Pacific Madrone (Arbutus menziesii* Pursh): *Current Theory and Research Directions.* Tacoma, WA: Pollard Group.

Hitchmough, J. 2004. "Philosophical and practical challenges to the design and management of plantings in urban greenspace in the 21st century." *Acta Horticulturae* 643:97–103.

Jurskis, V. 2005. "Eucalypt decline in Australia, and a general concept of tree decline and dieback." *Forest Ecology and Management* 215(1–3): 1–20.

Kendle, A. D., and J. E. Rose. 2000. "The aliens have landed! What are the justifications for 'native only' policies in landscape plantings?" *Landscape and Urban Planning* 47(1–2): 19–31.

Kuhns, M. R. 1998. "Urban/community forestry in the Intermountain West." *Journal of Arboriculture* 24(5): 280–85.

Morimoto, J., and H. Yoshida. 2005. "Dynamic changes of native Rhododendron colonies in the urban fringe of Kyoto city in Japan: Detecting the long-term dynamism for conservation of secondary nature." *Landscape and Urban Planning* 70(3–4): 195–204.

Muir, J. 1984. "Silvicultural information to help select and manage native trees in urban or suburban developments." *Arboricultural Journal* 8(1): 13–18.

Nash, L. J., and W. R. Graves. 1993. "Drought and flood stress effects on plant development and leaf water relations of five taxa of trees native to bottomland habitats." *Journal of the American Society for Horticultural Science* 118(6): 845–50.

Ware, G. H. 1980. "Little-known Asian elms: Urban tree possibilities." *Journal of Arboriculture* 6(8): 197–99.

Whitlow, T. H., and L. B. Anella. 1999. "Response of *Acer rubrum* L. (red maple) genotypes and cultivars to soil anaerobiosis." *Acta Horticulturae* 496:393–99.

Original article posted in September 2001.

GOING NATIVE? OR NOT?

While native plants are naturally adapted to their place of origin, it's also true that native plants did not evolve in cities. Urban areas usually have markedly different environmental conditions than the original landscape:

- Soils are heavily disturbed, often compacted, and bear little resemblance to native soils.
- Buildings, fences, and other hardscape structures reduce light availability to adjacent landscapes; in extreme cases only indirect light is available.
- Urban areas are nearly always warmer than surrounding sites. In addition, heat and/or light radiated from asphalt, concrete, and other surfaces can scorch plants not adapted to such extremes.
- Wind canyons are formed along streets with high-rises, retaining walls, or other structures.
- Heavy use of landscapes by vehicles, people, and animals will take its toll on soil and plant health.

Thus, urban landscapes are anything but natural and many native species will not thrive under such stressful conditions. In such cases, you might consider the benefits of nonnative species:

- Many nonnative species are not invasive; these include trees, shrubs, and perennials that have been part of our ornamental landscapes for decades (or centuries) and pose little risk to native species.
- There is often a wide selection of nonnative plants available; broaden your perspective to include species that are functionally and aesthetically appropriate for the landscape.
- Nonnative species can provide habitat for wildlife, especially in areas where native plants normally used by wildlife are rare or absent. Providing additional food and shelter for bats, birds, and insects will improve your landscape biodiversity and sustainability.

THE MYTH OF WELL-BEHAVED ORNAMENTALS

The Myth

"Garden plants do not become invasive."

We're all familiar with weeds in our landscapes. *Calystegia sepium* (hedge bindweed), *Equisetum arvense* (horsetail), *Taraxacum officinale* (dandelion), and *Cirsium arvense* and *Cirsium vulgare* (Canadian and bull thistle) are but a few of the weeds we battle in Pacific Northwest gardens. Larger herbaceous and woody perennials such as *Hedera helix* (English ivy), *Ulex europaeus* (gorse), *Cytisus scoparius* (Scots broom), *Rubus discolor* (Himalayan blackberry), and *Polygonum cuspidatum* (Japanese knotweed) are ubiquitous in parks and along roadsides. These species have cost countless hours of labor and gallons of herbicide in our quest to restore impacted landscapes to a more natural and diverse state.

Of course, we avoid this scenario in our own home landscapes;

few of us deliberately purchase weeds. Instead, we choose ornamental plants that are both attractive and tolerant of hostile environmental conditions. We desire plants that establish quickly, flower profusely, and spread so densely that weeds can't penetrate. These characteristics allow us to minimize labor and the use of pesticides, creating what appears to be a sustainable landscape. Isn't this a desirable outcome?

The Reality

Not all of our beloved garden plants behave themselves. The adjectives that epitomize some of our favorite choices—like "fast-spreading," "self-sowing," and "tolerates poor soil"—are also indicators of potential invasiveness. Some herbaceous weeds, and all of the woody ones listed above, were deliberately introduced into this country. Look at the common and Latin names; many of them indicate the country of origin. At one time they were seen as beneficial additions to our landscape in terms of ground cover (ivy), natural fences (gorse and Japanese knotweed), attractive flowers (Scots broom), or abundant fruit production (Himalayan blackberry). Unfortunately, these exotic visitors wore out their welcome by leaving the confines of the garden and spreading to other landscapes, particularly sensitive natural areas. In our remnant natural areas, these invaders quickly establish and eliminate many of the native flora, indirectly eliminating wildlife that depends on a diversity of plant life.

While few people would now purchase Scots broom (even if it were available), they do purchase other broom species because they have lovely flowers. Favorite shrubs and trees like *Buddleia davidii* (butterfly bush) and *Sorbus aucuparia* (European mountain ash) are already exhibiting bad manners and popping up along roadsides and in natural areas. Many invasive plants are readily available at nurseries across the country. Even species that have been deemed noxious weeds—English ivy in particular—are still available at nurseries, home improvement centers, and through the Internet. Should they be?

To address this issue on a national scale, individuals from diverse fields have collaborated to draft voluntary codes of conduct for government agencies, nursery professionals, landscape architects, botanic gardens and arboreta, and the gardening public. This effort is being supported by a broad array of organizations representing the nursery industry (American Nursery & Landscape Association), botanical gardens (American Association of Botanical Gardens and Arboreta), landscape architects (American Society of Landscape Architects), scientists (USDA Agricultural Research Service), and conservationists (Center for Plant Conservation). Through self-governance and self-regulation, these codes encourage behaviors that detect and prevent future introductions of invasive ornamental species to North America.

The Bottom Line

Here are some suggested ways that nursery professionals and the gardening public can participate in regional and national efforts to reduce the effect of invasive plants:

- Phase out existing stocks of regionally invasive species.
- Purchase and promote noninvasive, environmentally safe plant species.
- Remove invasive species from your land and replace them

with noninvasive species suited to site conditions and landscape usage.

- Work with neighbors or volunteer at botanical gardens and work parties in natural areas to eliminate populations of invasive plants.

- Visit the Web site at www.mobot.org/iss to learn more about preventing plant invasions.

References

Batianoff, G., and A. J. Franks. 1997. "Invasion of sandy beachfronts by ornamental plant species in Queensland." *Plant Protection Quarterly* 12(4): 180–86.

Blood, K. 1999. "Garden plants under the spotlight." *Plant Protection Quarterly* 14(3): 103.

Boersma, P. D., S. H. Reichard, and A. N. Van Buren, eds. 2006. *Invasive Species in the Pacific Northwest.* Seattle, WA: University of Washington Press.

Kay, S. H., and S. T. Hoyle. 2001. "Mail order, the Internet, and invasive aquatic weeds." *Journal of Aquatic Plant Management* 39:88–91.

Meyer, J. Y., and C. Lavergne. 2004. "Beautes fatales: Acanthaceae species as invasive alien plants on tropical Indo-Pacific islands." *Diversity and Distributions* 10(5–6): 333–47.

Pennucci, A., and T. Griffin. 1998. "The control of escaped garden plants in New England cemeteries." *Proceedings of the 52nd Annual Meeting of the Northeastern Weed Science Society,* Geneva, New York, 61–65.

Peters, W. L., M. H. Meyer, and N. O. Anderson. 2006. "Minnesota horticultural industry survey on invasive plants." *Euphytica* 148(1–2): 75–86.

Shaw, R. 2003. "Aliens on the march." *The Garden—Journal of the Royal Horticultural Society* 128(6): 464–65.

Siepen, G. L., J. Morton, and P. Donatiu. 2003. "Weeds and garden magazines." *Ecological Management and Restoration* 4(2): 153–55.

Snow, K. 2001. "Working with the horticultural industry to limit invasive species introductions." *Conservation Biology in Practice* 3(1): 33–36

USDA Natural Resources Conservation Service. Invasive and Noxious Weed Lists. http://plants.usda.gov/java/noxiousDriver. Accessed April 5, 2007.

Original article posted in May 2002.

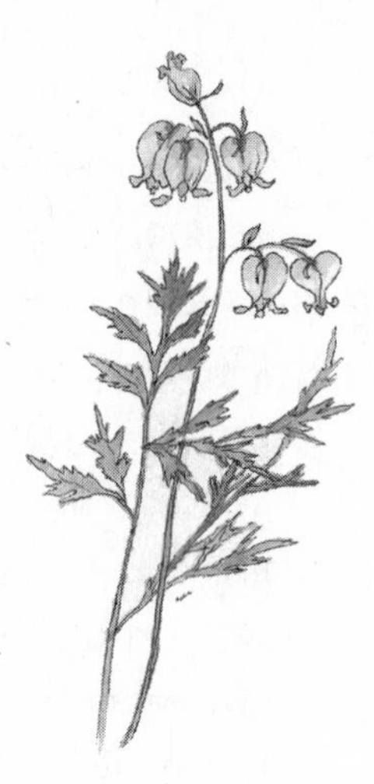

THE MYTH OF PLANT QUALITY

The Myth

"The quality of plant material is directly proportional to the price."

One month I was having difficulty deciding on a topic for my column. To stimulate my creative juices, I tackled our home landscape-in-progress and ended up shopping for plants over two weekends. After spending several hundred dollars and installing my new shrubs and trees, I had identified a topic that, to put it mildly, I felt passionate about. I guess you could call this myth "You get what you pay for." The working hypothesis is that pricey retail nurseries are extra careful with their purchasing and care of plant materials, and therefore consumers will find the best quality stock at these nurseries.

The Reality

Since I was installing a variety of shrubs and trees, I limited my shopping to three well-known and respected nurseries with wide plant selections. At the first nursery I inspected some *Ginkgo biloba* cultivars. They were six to seven feet tall and nicely shaped—but in two-gallon containers that were obviously too small for trees of this size. When I pulled one out of the pot, I found finger-sized woody roots completely encircling the root ball. Eighty dollars for this? I don't think so. All of the *Ginkgo* cultivars had the same fatal root problem.

At the next nursery, I found a white flowering *Ribes sanguineum*. It was relatively loose in the soil, so I knew it had been recently potted up, but still appeared to be a healthy specimen. When I pulled this three-foot shrub out of the pot at home, I found a four-inch cubical, woody root mass with new fibrous roots growing from it. Out came the pruners, and away went most of the roots. This plant will need extra care to ensure that it establishes on such a reduced root system.

At the third nursery (now I was really being careful), I looked for a *Cercis canadensis*. I avoided the balled and burlapped plants since the balls were all too small in comparison to the crown size, and inspected the containerized stock. They were firmly rooted, and I almost bought one. Then I looked at the crown more carefully and found that all of the leaders had been pruned back to give the plant the appearance of a miniature adult specimen. Unless these trees are correctively pruned and a new leader is established, they will never regain their characteristic form. Disgusted, I put back the plant and settled on five-gallon containers of *Acer circinatum* and *Philadelphus lewisii*. Both had nice crown shapes and were relatively loose in the soil—maybe a bit too loose.

I was getting desperate, however, and needed to get my gardening fix. So I bought them at premium prices. At home I discovered that they were simply bare-root stock covered with new potting media. These bare-root plants consisted of only the woody storage roots and had none of the fine, flexible, fibrous roots needed for water and nutrient uptake. They will survive with extra care, but I resented paying for five-gallon containers of potting media.

I could have taken all of these plants back, but what would I have purchased instead? Every single shrub and tree I bought had some serious root problem, and I'd avoided many other plants with more obvious problems. It takes time to return plants, and it is especially difficult in metropolitan areas, where driving is everyone's most hated activity. So I kept them all, and they will receive extra TLC. But in a year when drought is looming on the horizon, this is not an environmentally sound practice.

As I've discussed previously, circling, girdling, or kinking roots begin when plants are left too long in small containers. If uncorrected, these root flaws will lead to the early decline and death of trees and shrubs. It behooves the nursery and landscape industry to take an active role in addressing this problem through improved propagation practices and more critical inspection of plant materials. Consumers are becoming better educated, and to retain their trust, nurseries need to be more responsible for the quality of material they sell.

The Bottom Line

- Most consumers are not careful in their selection of quality plant material, but even knowledgeable consumers can miss serious root problems at the retail nursery. Even if they are careful in selecting healthy plants, consumers must realize that potentially fatal root flaws may not be discovered until transplanting.

- Retail nurseries need to be more vigilant in their inspection and acceptance of containerized and balled and burlapped materials.

- Production nurseries need to modify potting-up practices to ensure that root balls are properly trained.

- All nurseries need to end the practice of improperly pruning young trees to mimic adult form.

References

Chalker-Scott, L. 2005. "Comparisons of plant quality among large production nurseries." Final report to Washington State Department of Agriculture. http://agr.wa.gov/PlantsInsects/NurseryInspection/NurseryResearch/NurseryResearch.htm. Accessed May 5, 2007.

Gingras, B. M., S. Richard, and N. Robert. 2002. "Comparative performance after five years of large-sized conifers and broadleaves grown in air-slit containers." *Mémoire de recherche forestière* 141. *Direction de la recherche forestière, Forêt Québec,* QC.

Lovelace, W. "Root production method system." 2002. *USDA Forest Service RMRS* P24:20–21.

Srivastava, A., T. S. Rathore, G. Joshi, and K. S. Reddy. 2002. "Modern nursery practices in the production of quality seedlings of important forestry species using root trainer technology." *Myforest* 38(3): 257–63.

Original article posted in May 2001.

CHOOSING THE BEST PLANTS AT THE NURSERY: THE BIG PICTURE

- Do trees have heights appropriate to pot size? Ideally, trees are potted up rapidly to accommodate their expanding root systems. Tall trees in small pots will probably have poor roots.
- Do trees have a strong central leader? (This does not apply to multi-trunked species). Be sure that there is no evidence of topping, which will create competing leaders and poor crown structure. Topped trees will require constant maintenance unless their crown structure is restored through proper pruning.
- Is there evidence that side branches have been removed? If so, the tender bark of these trees may be susceptible to sunscald and other damage. Evidence of branch removal is easily seen through the presence of limb scars along the trunk.
- Do trees have good taper? Good taper is evident if the diameter at the base of the tree is greater than diameter closer to the crown.
- Are trees tightly staked? If so, they probably will have poor taper and may suffer crown breakage once stakes are removed. Staking, if present, should allow the crown to move in the wind.

Continued on next page

CHOOSING THE BEST PLANTS AT THE NURSERY: UP CLOSE AND PERSONAL

- Are there poor branch angles that will result in included bark? Generally, you will want to avoid trees with branches that form acute angles with the main trunk. Such branches are more likely to break the larger they become.
- Are suckers growing from the base of the tree? This can indicate a poor attachment in grafted specimens or an impaired root system in nongrafted trees. (This does not apply to multitrunked species.)
- Is the root crown visible? This region of the tree, also called the root flare, is where the trunk merges with the root system. Ideally, you should avoid trees and shrubs whose root crown is buried or covered with burlap, as this creates a welcome environment for pests or disease. It is acceptable to gently brush soil or peel back burlap from the trunk to inspect the root crown. If the root crown is buried too deeply for you to find, look for another specimen.
- Are there multiple layers of materials around the roots—for instance, burlap, wire baskets, and containers all together? These trees are harder to inspect and are often plants that have been in the nursery for a while.
- Are there roots on the soil surface—especially circling or hooked roots? This indicates serious root problems below ground.
- Does the trunk appear abnormal in any way, especially near the root crown or on grafted material? This may be evidence of damage or disease. Look for better specimens without ready-made problems.

THE MYTH OF DRAINAGE MATERIAL IN CONTAINERS

The Myth

"Add a layer of gravel or other coarse material in the bottom of containers to improve drainage."

This is just one of those myths that refuses to die, regardless of solid scientific evidence to the contrary! Nearly every book or Web site on container gardening recommends placing coarse material at the bottom of containers for drainage. The materials most often recommended for this practice are sand, gravel, pebbles, and pot shards. Other "benefits" often mentioned include preventing creatures from entering through the drain holes and stabilizing the container.

Some of these recommendations are quite specific and scientific sounding. Consider this advice from a 1960s book on container plants: "Adequate drainage is secured by covering the hole in the bottom of the pot with a piece of broken flowerpot, concave side down; this in turn is covered with a layer (½" to 1" deep) of flowerpot chips. On top of this, a ¼" to ⅜" layer of coarse organic material, such as flaky leaf mold, is placed." The advice seems to make perfect sense, and it's presented so precisely. After all, we know that plants need good drainage so their roots will receive adequate oxygen, and we also know that water passes through coarsely textured material faster than it does fine material. So what's not to like?

The Reality

Nearly 100 years ago, soil scientists demonstrated that water does not move easily from layers of finely textured material to layers of more coarsely textured material. Since then, similar studies have produced the same results. One study found that more moisture was retained in soil underlain by gravel than in that underlain by sand. Therefore, the coarser the underlying material, the more difficult it is for water to move across the interface. Imagine what happens in a container lined with pot shards!

Gravitational water will not move from a finely textured soil into a coarser material until the finer soil is saturated. Since the stated goal for using coarse material in the bottoms of containers is to "keep soil from getting water logged," it is ironic that adding this material will induce the very state it is intended to prevent.

The Bottom Line

- "Drainage material" added to containers will only hinder water movement.

- Planting containers must have drainage holes for root aeration.

- Use the same planting material throughout the entire container to ensure optimal water and air movement.

References

Ju, S. H., and K. J. S. Kung. 1997. "Steady-state funnel flow: Its characteristics and impact on modeling." *Soil Science Society of America Journal* 61(2): 416–27.

Kelling, K. A., and A. E. Peterson. 1975. "Urban lawn infiltration rates and fertilizer runoff losses under simulated rainfall." *Proceedings of the Soil Science Society of America* 39(2): 348–52.

Kim, J. G., C. M. Chon, and J. S. Lee. 2004. "Effect of structure and texture on infiltration flow pattern during flood irrigation." *Environmental Geology* 46(6–7): 962–69.

Kung, K. J. S. 1993. "Laboratory observation of funnel flow mechanism and its influence on solute transport." *Journal of Environmental Quality* 22(1): 91–102.

Nektarios, P. A. 2004. "Visualization of preferential flow in simulated golf course putting green profiles." *Rasen Turf Gazon* 35(4): 56–60.

Schaetzl, R. J. 1996. "Spodosol-Alfisol intergrades: Bisequal soils in NE Michigan, USA." *Geoderma* 74(1–2): 23–47.

Spomer, L. A. 1990. "Evaluating 'drainage' in container and other shallow-drained horticultural soils." *Communications in Soil Science and Plant Analysis* 21(3–4): 221–35.

Warrick, A. W., P. J. Wierenga, and L. Pan. 1997. "Downward water flow through sloping layers in the vadose zone: Analytical solutions for diversions." *Journal of Hydrology* 192(1–4): 321–37.

Yoder, R. E. 2001. "Field-scale preferential flow at textural discontinuities. *Proceedings of the 2nd International Symposium on Preferential Flow: Water Movement and Chemical Transport in the Environment*, American Society of Agricultural Engineers, St. Joseph, Michigan, 65–68.

Original article posted in November 2000.

THE MYTH OF COLLAPSING ROOT BALLS

The Myth

"Balled and burlapped root balls must be left intact during transplanting."

While shopping for trees at my favorite nursery, I recently overheard a customer ask a staff person about installing her newly purchased balled and burlapped tree. "When I plant my tree, I should take off the burlap and twine, right?" she asked. "Oh no!" exclaimed the staffer. "You don't want to disturb the root ball. Just peel the burlap back from the trunk and leave the rest intact. Otherwise, the root ball will collapse and the tree will die."

At first glance, this appears to be reasonable advice. Balled and burlapped trees are much heavier than containerized plants, and one can visualize the unbound root ball collapsing and crushing the root system. The weight of the root ball also helps stabilize the tree and prevent tilting or falling. Finally, the root ball soil contains beneficial microbes and other soil organisms that can help ease transplant shock to the root system. With these benefits in mind, why would you consider removing the burlap?

The Reality

Many nurseries will not guarantee their plant materials if the customer disturbs the root ball (because exposed roots require extra care), so customers are loath to do anything that might negate the guarantee. This is unfortunate, as disturbing the root ball is exactly what you want to do to maximize survival of your newly transplanted tree.

The most important reason to disturb the root ball of a balled and burlapped tree is to inspect the root system. The circling, girdling, kinked, and hooked root systems often found in containerized plants occur frequently in balled and burlapped plants, too. Nearly every balled and burlapped tree I have purchased and installed, either in my own landscape or as part of a project, has had serious root defects. By removing the heavy clay one can find and correct many of these defects. Without corrective pruning these defects will significantly lower the life span of your tree. Remember, root pruning stimulates the growth and development of new roots that will enhance tree establishment in the landscape.

A second reason to break up the root ball is to remove the clay soil that makes the tree so heavy in the first place. Most balled and burlapped trees are grown in a soil with clay characteristics so that when the tree is dug the root ball will hold its shape; sandy soil would simply fall away from the roots. The clay soil not only maintains its shape but also retains water, so that balled and burlapped plants are usually more hydrated during the time they are out of the ground. When the tree is planted into the landscape, however, the clay character of the soil is often different from that of the surrounding native soil. Differences between soil textures will impede water movement and therefore inhibit root establishment.

A final reason to remove the bagging materials and root ball soil is that many of the balled and burlapped specimens at the nursery have been burlapped too high during field digging and bagging. Burlap and soil that cover the trunk above the root crown will lead to trunk disease and death. In every nursery I've visited I have found more than one tree trunk literally rotting in the bag. Before purchasing any balled and burlapped stock you should ensure that a healthy trunk lives beneath the burlap.

The best practice for transplanting balled and burlapped trees is relatively straightforward:

1 Remove all wire baskets, twine, and burlap from the root ball. Working on top of a tarp will allow you to transport the root ball remnants elsewhere.

2 Remove all clay from the root ball. This can be done most easily by using a water bath or a hose. Use your fingers to work out clumps of clay from between the roots. Save the water.

3 Look for and prune out defects in your freshly denuded roots. Be sure to keep the roots moist during this procedure and work in the shade, if possible.

4 Dig the planting hole to be as deep as the root system and at least twice as wide. The hole will resemble a shallow bowl.

5 Form a soil mound in the center of the hole to support the root crown of the tree, set the tree on top of the mound, and arrange the roots radially.

6 Backfill with native soil; do not use any type of soil amendment. Do not step on the root zone, but gently firm using your hands.

7 Water in well, preferably using the water from step 2, which will contain nutrients and microbes. Add an appropriate fertilizer (i.e., primarily nitrogen and little or no phosphorus).

8 Mulch the entire planting region with at least four inches of organic mulch, keeping a buffer between the trunk and the mulch, to prevent disease.

9 If necessary, stake your tree low and loose with three stakes for no longer than one year after planting. The watering-in process may make staking unnecessary, as a thick, solid mud will encase the roots and prevent their movement.

10 Keep your tree well-watered during the first year of establishment. You may have removed a good portion of the root system, and its ability to take up water and nutrients will be temporarily impaired. Do not succumb to the temptation to crown prune or add expensive, but pointless, transplant supplements.

This method is radically different from historically accepted practices. Yet recent and ongoing research demonstrates that bare-rooting balled and burlapped trees leads to substantial increases in tree establishment and survival. Investing the time to prepare and install trees properly will pay future dividends of reduced maintenance and mortality for the lifetime of your landscape.

The Bottom Line

- Root defects can only be found and corrected if root ball soil is removed.
- Balled and burlapped plant materials usually contain soil significantly different from that of the transplant site.
- Differences in soil texture will impede both water movement and root establishment.
- Proper root preparation, combined with good installation techniques, will greatly improve tree establishment and survival in any landscape.

References

Cole, J. C., and D. L. Hensley. 1994. "Field-grow fabric containers do not affect transplant survival or establishment of green ash." *Journal of Arboriculture* 20(2): 120–23.

Davidson, H., and W. Barrick. 1975. "Fertilizer practices and vegetation control on highway landscape plantings." *Farm Science Series Research Report* 262.

Grover, B. L., G. A. Cahoon, and C. W. Hotchkiss. 1964. "Moisture relations of soil inclusions of a texture different from the surrounding soil." *Proceedings of the Soil Science Society of America* 28: 692–95.

Harris, J. R., P. Knight, and J. Fanelli. 1996. "Fall transplanting improves establishment of balled and burlapped fringe tree (*Chionanthus virginicus* L.)." *HortScience* 31(7): 1143–45.

Heisler, G. M ., R. E. Schutzki, R. P. Zisa, H. G. Halverson, and B. A. Hamilton. 1982. "Effect of planting procedures on initial growth of *Acer rubrum* L. and *Fraxinus pennsylvanicum* L. in a parking lot." USDA Forest-Service research paper, NE-513.

Hensley, D. L. 1993. "Harvest method has no influence on growth of transplanted green ash." *Journal of Arboriculture* 19(6): 379–82.

Kuhns, M. R. 1997. "Penetration of treated and untreated burlap by roots of balled-and-burlapped Norway maples." *Journal of Arboriculture* 23(1): 1–7.

Smith, E. M. 1971. "Evaluating the lasting quality of burlap and containers." *Ohio Agricultural Research and Development Center Research Summary* 56: 59–60.

Original article posted in January 2003.

THE MYTH OF TREE STAKING

The Myth

"Newly planted trees should be staked firmly and securely."

When I moved back to Washington State in 1997, I was appalled to see tree bondage rampant in many urban landscapes. The oozing, swollen wounds around staking wires are just too much for me to bear, and I will admit to playing tree liberator on more than one occasion.

Like planting hole amendment, tree staking is done with the best of intentions but without regard to long-term tree health. Rather than helping a tree develop the root and trunk growth that allow it to stand independently, improper tree staking replaces a supportive trunk and root system. This artificial support causes the tree to put its resources into growing taller but not growing wider. When the stakes are removed (if they ever are), the lack of trunk and root development makes these trees prime candidates for breakage or blowdown. A comparative example is what is seen when a forest is cleared for a housing development. A few trees near the center of the stand are left on the lots; these trees are tall and skinny, with well-developed crowns. But in the first good windstorm, down these trees come. They have lost the supportive protection of the surrounding trees and are unable to stand alone.

It's interesting (and comforting) to find that nearly all current books and reputable Web pages agree with this assessment of staking. Then why are there so many incorrectly staked trees in the landscape? I believe that there are several contributing factors. First, containerized nursery materials are often staked for temporary stability, because these plants are often top-heavy, with large crowns and small pots. Many consumers don't understand that the staking material needs to be removed when the tree is transplanted. Moreover, the oral and written information from some retail nursery centers instruct their customers to stake their trees whether or not there is a need for doing so. These instructions are sometimes incorrect in addition to being unnecessary. This misinformation is also spread by professionals, including landscape architects, who use outdated (and incorrect) specifications for staking procedures. Landscape contractors must follow these specifications, but more often than not there is no provision for removing the stakes after the trees have established. Thus, we can find trees incorrectly staked in public and commercial sites as well as home landscapes.

The first two practices are probably responsible for most of the incorrect staking in home landscapes, while the last two factors are probably responsible for most of the incorrect staking in public and commercial landscapes.

The Reality

The three cardinal sins of tree staking are:

- staking too high (greater than two-thirds of tree height)
- staking too tightly (materials must be flexible and allow the tree to move)
- staking too long (all materials must be removed after one year)

Trees that are staked improperly will:

- grow taller, but with decreased trunk caliper
- develop less trunk taper (or even a reverse trunk taper)
- develop xylem unevenly
- develop a smaller root system
- suffer rubbing and girdling injuries from stakes and ties
- be more likely to snap in a high wind after stakes are removed
- often be unable to remain upright after stakes are removed

If it is necessary to stake a tree, there are acceptable methods that allow proper trunk and root development while providing temporary support and protection. This is especially important in urban areas because of poor, shallow soils that hinder root development and the potential of mechanical injury from people and vehicles. (In fact, properly installed tree guards made of decorative grillwork are an excellent way of protecting street trees permanently from mechanical damage. These tree guards are not attached to the trees but stand alone and should be upsized as the tree grows.)

The Bottom Line

- Most containerized and balled and burlapped materials do not need staking; bare-root trees often do.
- If trees must be staked, place stakes as low as possible but no higher than two-thirds the height of the tree.
- Materials used to tie the tree to the stake should be flexible and allow for movement all the way down to the ground so that trunk taper develops correctly.
- Remove all staking material after roots have established. This can be as early as a few months, but should be no longer than one growing season.
- Materials used for permanent tree protection, such as tree guards and gratings, should never be attached to the tree.

References

Harris, R. W. 1984. "Effects of pruning and staking on landscape trees." *Journal of Environmental Horticulture* 2(4): 140–42.

Harris, R. W., and W. D. Hamilton. 1969. "Staking and pruning young *Myoporum laetum* trees." *Journal of the American Society for Horticultural Science* 94:359–61.

Leiser, A. T., R. W. Harris, P. L. Neel, D. Long, N. W. Stice, and R. G. Maire. 1972. "Staking and pruning influence trunk development of young trees." *Journal of the American Society for Horticultural Science* 97(4): 498–503.

Leiser, A. T., and J. D. Kemper. 1973. "Analysis of stress distribution in the sapling tree trunk." *Journal of the American Society for Horticultural Science* 98(2): 164–70.

Leiser, A. T., and J. D. Kemper. 1968. "A theoretical analysis of a critical height of staking landscape trees." *Proceedings of the American Society for Horticultural Science* 92:713–20.

Patch, D. 1981. "Tree staking." *The Garden—Journal of the Royal Horticultural Society* 106(12): 500–503.

Telewski, F. W., and M. L. Pruyn. 1998. "Thigmomorphogenesis: A dose response to flexing in *Ulmus americana* seedlings." *Tree Physiology* 18(1): 65–68.

Whalley, D. 1982. "Lighter the better [tree staking]." *Gardener's Chronicle and Horticultural Trade Journal* 191(24): 23–24.

Wrigley, M. P., and G. S. Smith. 1978. "Staking and pruning effects on trunk and root development of four ornamental trees." *New Zealand Journal of Experimental Agriculture* 6(4): 309–11.

Original article posted in March 2001.

SOIL ADDITIVES

THE MYTH OF SOIL AMENDMENTS, PART 1

The Myth

"When transplanting trees or shrubs into landscapes, amend the backfill soil with organic matter."

Of all the fictions that abound in popular horticulture, none is as deceptive as this one. It stems from the old adage "to dig a five-dollar hole for a fifty-cent plant." It seems logical that steer manure, peat moss, compost, et cetera, would improve poor soils by increasing aeration, nutritional value, and water-holding capacity. And it does—in the immediate vicinity of the planting hole. But eventually, amending planting holes will have negative consequences on plant health. To understand why, it's necessary to examine plant physiology and soil-water relations.

The Reality

Let's say you have incorporated the recommended 25–50 percent organic matter into your backfill. (Remember that an ideal soil contains 5 percent organic matter by volume.) The initial results are positive; roots grow vigorously in this ideal environment as long as irrigation is provided. But what happens when these roots encounter the interface between the planting hole and the native soil? Native soil contains fewer available nutrients, is more finely textured, and is less aerated. Roots react much in the same way as they do in containers: They circle the edge of the interface and grow back into the more hospitable environment of the planting hole. The roots do not establish in the native soil, eventually resulting in reduced growth rates and increased hazard potential as crown growth exceeds root-ball diameter.

Soil-water movement is problematic as well. Amended backfill has markedly different characteristics from the surrounding native soil; it is more porous, and water will wick away to the finer-textured native soil. In the summer, moisture within the planting hole will be depleted but not replaced by the water held more tightly in the native soil. This results in water stress to the plant unless the planting hole is kept irrigated, a costly and often unrealistic practice. During wet seasons water will move quickly through the amended soil only to be held back by the more slowly draining native soil. The resulting bathtub effect, wherein water accumulates in the planting hole, floods the roots and eventually kills the plant.

Finally, all organic material eventually decomposes. If you've incorporated one quarter or one half organic matter by volume to the planting hole, within a few years you will have a sunken garden in your landscape. This only exacerbates the flooding problem during wet conditions.

No long-term scientific studies to date show any measurable

benefit of soil amendment, except in containerized plant production. Plants grown in native soil have consistently shown better root establishment and more vigorous growth. Only one study reported no negative effects of amending soil with organic matter—but there were no benefits, either. When you consider the cost of materials and labor needed to incorporate soil amendments, it's difficult to justify the practice.

This outdated practice is still required in the specifications of architects, landscapers, and other groups associated with landscape installation. It is still recommended by garden centers and gardening articles. And there is a multimillion-dollar soil-amendment industry that has little interest in debunking this myth. As responsible green gardeners, we need to recognize and avoid nonsustainable management practices.

The Bottom Line

- Don't be an "enabler": Use native soils for backfill without amendment.
- Amended planting holes inhibit water movement between the soil in the hole and that in the surrounding landscape.
- Amended planting holes inhibit root exploration into the surrounding soil, reducing plant establishment and survival.
- Eventually, plants in amended planting holes will sink below grade as organic material decomposes and soil settles.
- Instead of amending the planting hole, add organic material as a topdressing—a practice that is both easy and sustainable.

References

Corley, W. L. 1984. "Re-evaluating the value of amending planting holes with organic material." *American Nurseryman* 159(6): 113–16.

Gilman, E. F. 2004. "Effects of amendments, soil additives, and irrigation on tree survival and growth." *Journal of Arboriculture* 30(5): 301–10.

Smalley, T. J., and C. B. Wood. 1995. "Effect of backfill amendment on growth of red maple." *Journal of Arboriculture* 21(5): 247–50.

See also references in "The Myth of Soil Amendments, Part 2" and "The Myth of Soil Amendments, Part 3."

Original article posted in August 2000.

THE MYTH OF SOIL AMENDMENTS, PART 2

The Myth

"If you have a clay soil, add sand to improve its texture."

I was waffling on what myth to debunk one month when I received an issue of *American Nurseryman*. One of the articles featured a high-end landscape renovation. The horticulturist on this job stated, "The soil was bad. . . . We had to completely renovate the soil types with compost and sand."

The Reality

As I've already dealt with the fallacy of incorporating organic amendments to permanent landscape installations, I won't repeat it here. But the equally misguided practice of adding sand to improve a clay soil texture needs to be addressed. Soil texture is determined by particle size, which ranges from microscopic clay flakes to rounded silt particles to sand grains. While undisturbed sandy soils are well aerated and well drained, they are nutrient poor since sand and silt cannot bind mineral nutrients. In contrast, clay soils do bind mineral nutrients but have poor drainage and aeration. Thus, a soil with both sandy and clay characteristics should be optimal for plant root health. So it's easy to see how the practice of adding sand to clay soils has evolved.

The problems occur when sand and clay are mixed in incorrect proportions. An ideal soil has 50 percent pore space (or "holes" between soil particles), with the remaining 50 percent consisting of minerals and organic matter. The pore spaces in a clay soil are all small, while those in a sandy soil are all large. When one mixes a sandy and a clay soil together, the large pore spaces of the sandy soil are filled with the smaller clay particles. This results in a heavier, denser soil with less total pore space than either the sandy or the clay soil alone. (A good analogy is the manufacture of concrete, which entails mixing sand with cement—a fine particle substance. The results are obvious.) A soil must consist of nearly 50 percent sand by total volume before it takes on the characteristics of a sandy soil. For most sites, it would be prohibitively expensive to remove half the existing soil and add an equal volume of sand and then till it to the necessary eighteen- to twenty-four-inch depth. Mineral amendments of large-particle size, such as perlite, may provide some benefit, but can also be costly depending on the size of the site. (Reducing this task to amending only the planting hole is a recipe for plant failure, as the previous chapter has shown.)

The relative percentages of clay, silt, and sand in a soil will help determine its structure. Predominantly sandy and silty soils don't have much structure—they tend to erode easily, like sand castles at the beach. Soils with more clay content, such as the various loams, aggregate into larger chunks called "peds." (Heavy clay soils, however, have a structure better suited for throwing pots than for growing many landscape plants.) Highly aggregated soils are optimal for root growth and aeration, but can be easily destroyed by any activity that results in soil compaction.

Soil structure can be improved through proper site preparation and management. One of the least invasive and most cost-effective ways to do this is through the use of organic mulches. Organic mulches are especially effective in protecting soil structure in landscapes that receive high-volume foot traffic. My landscape restoration classes now routinely spread wood chips on the site to allow soil recovery to begin as they prepare the site and install new plants. One particular site, a small lot near a bus stop, consisted of weeds, bare soil, and a few existing trees and shrubs. When we tried to take a soil core, the corer bent! We had eight to ten inches of wood chips spread over the whole site. A month later, we moved aside part of the mulch and dug out a shovelful of rich, loamy soil. Had I not seen it for myself, I'm not sure I would have believed these stunning results. The addition of the wood chips allowed the site to retain soil moisture and reduced the constant impact from foot traffic, thus enabling the soil to regain its structure.

The Bottom Line

- All good landscape soils must contain some clay for optimal plant growth.

- Many urban soils are compacted and are mistaken for heavy clay soils.

- The structure of heavily compacted soils can be improved through the use of protective mulches.

- Many problems associated with clay soils (poor aeration, drainage, et cetera) can be alleviated through selecting appropriate plants and protecting soil from compaction.

References

Spomer, L. A. 1983. "Physical amendment of landscape soils." *Journal of Environmental Horticulture* 1(3): 77–80.

See also references in "The Myth of Soil Amendments, Part 1" and "The Myth of Soil Amendments, Part 3."

Original article posted in November 2000.

THE MYTH OF SOIL AMENDMENTS, PART 3

The Myth

"Healthy soil has high organic content."

One of the newer catchphrases making the rounds in print and on gardening Web sites is "building healthy soil." The starting premise is that residential soil is inherently unhealthy and in need of amendment. Without fail, these media sources recommend the incorporation of large volumes of compost as a means of improving soil structure, adding nutrients, improving drainage and aeration, and increasing water-holding capacity. Web sites recommend adding anywhere from one to four inches of compost; one site suggests using "1 part compost to 2 parts dirt." In general, the message is that unamended soil is unacceptable, and the only way to make it healthy is by adding large quantities of compost.

The Reality

The dubious practice of amending soil areas destined for permanent landscape installations has been discussed in the previous chapters. To summarize briefly, the problem with this practice is that within ten years (conservatively) the organic amendment will have decomposed; one is then left with the original soil, which will have subsided and compacted during this time. You can see evidence of this practice by looking at older residential lawns: The lawns slope away from sidewalks and driveways and are inches below the grade of surrounding surfaces. There is no way to incorporate additional amendments into these permanent landscapes without damaging root systems. Instead, it is easier, cheaper, and more natural to add organic material by top-dressing landscapes that are not planted and harvested on an annual basis. (My fondness for wood chip mulches has been expressed before!)

What about landscapes that are planted and harvested on an annual basis—including vegetable gardens and flower beds? These landscapes are more logically managed by agricultural models—adding organic matter to replace nutrients removed from the soil by flowers and vegetables. The annual incorporation of compost makes sense here. However, one needs to have an idea of what the soil already contains before more material is added.

During home construction, topsoil is removed from the site and eventually replaced by "designed soil." It is almost impossible to purchase native topsoil in urban areas; it is too precious a commodity. Commercially available topsoil is usually a mixture of native topsoil and a variety of inorganic and organic materials, including sand, perlite, compost, peat moss, bark, sawdust, and manure. These designed soils usually contain 15 percent organic matter by weight (equivalent to 30 percent compost by volume). By comparison, native topsoils in Western Washington contain about 5 percent organic matter by weight (or 10 percent organic matter by volume); this level of organic matter is considered to be optimal in terms of nutrient content. Obviously, new residential landscapes contain high levels of organic matter, well above what is considered ideal.

If you don't know what your soil already contains in terms of nutrients, how can you possibly determine how much organic matter to add? It is simple and cheap to have your soil tested for organic matter content and nutrient levels, and this should be done at least once to determine baseline values. This information can help you determine if you need to add more organic material, and which nutrients in particular are at minimal levels. It wastes resources, both financial and natural, to add excessive amounts of organic matter without these baseline values.

One fall, my class collected soil samples from a local organic demonstration garden and sent them out for nutrient analysis; the garden had been experiencing some soil and plant-health problems. Every single one of the samples that was tested came back with nutrient readings off the scale. In large capital letters the report warned, "DO NOT FERTILIZE THIS SOIL." The excessive addition of nutrient-rich compost to this landscape contributed not only to plant-health problems by creating mineral toxicities but also to nutrient loading of adjacent natural waters.

The Bottom Line

- Do not incorporate organic amendments into landscapes destined for permanent installations; top-dress with mulch instead.

- Ideal soils, from a fertility standpoint, are generally defined as containing no more than 5 percent organic matter by weight, or 10 percent by volume (in western Washington).

- Before you add organic amendments to your garden, have your soil tested to determine its organic content and nutrient levels.

- Be conservative with organic amendments; add only what is necessary to correct deficiencies and maintain organic matter at ideal levels.

- Abnormally high levels of nutrients can have negative effects on plant and soil health.

- Any nutrients not immediately utilized by microbes or plants contribute to nonpoint source pollution.

References

Brady, N. C., and R. Weil. 1999. *The Nature and Properties of Soils*, 12th ed. Upper Saddle River, NJ: Prentice-Hall.

Craul, P. J. 1992. *Urban Soil in Landscape Design*. New York, NY: Wiley.

Day, R. W. 1994. "Performance of fill that contains organic matter." *Journal of Performance of Constructed Facilities, ASCE* 8(4): 264–73.

See also references in "The Myth of Soil Amendments, Part 1" and "The Myth of Soil Amendments, Part 2."

Original article posted in March 2003.

THE MYTH OF PHOSPHATE FERTILIZER, PART 1

The Myth

"Phosphate fertilizers will stimulate root growth of transplanted trees and shrubs."

This commonly spread myth originates from the legitimate addition of phosphorus to agricultural fields. Phosphorus is one of the inorganic macronutrients needed by all plants for the manufacture of phosphate-containing nucleic acids, ATP, and membrane lipids. Soils that have been heavily used for agricultural crops are often deficient in phosphorus, as are acid sandy and granitic soils. In landscaped urban soils, however, phosphorus is rarely deficient, and the misapplication of this element can have serious repercussions on the plant, the soil environment, and adjoining watersheds.

The Reality

When an element is deficient in the soil, it is considered limiting, and plant growth slows. This phenomenon is called "environmental dormancy." When the deficient element is added, the environmental constraint is lifted, and plant growth resumes at the normal rate if nothing else is limiting. Somehow the observation of growth restoration has been interpreted as growth stimulation (i.e., a growth rate greater than normal), and hence fertilizers are often regarded as miraculous compounds (just look at the names of some of them!).

One of the classic symptoms of phosphorus deficiency is reddening of the leaves. Unfortunately, many environmental stresses also induce foliar reddening; examples include cold temperature, high light intensity, insect damage, and drought. Urban landscape plants are much more likely to experience one of these stresses than phosphate deficiency.

Nitrogen is much more likely to be limiting in urban landscapes than phosphorous and can often improve the establishment and survival of transplanted trees and shrubs. Nitrogen deficiency is characterized by uniform leaf chlorosis (yellowing). Among other things, a lack of nitrogen reduces a plant's ability to take up phosphorus. When nitrogen is restored to optimal levels, the plant's ability to scavenge phosphorus from the soil is markedly improved. It's important to realize that when nitrogen is deficient, it does not logically follow that other nutrients must be deficient as well.

Because nitrogen is so often deficient in an actively growing landscape, the addition of ammonium sulfate (or some other nitrogen source) usually restores shoot growth. Phosphate addition, on the other hand, often has no apparent effect on shoots (probably because it's generally not limiting in perennial landscapes). This observation has led landscapers and fertilizer manufacturers to claim that phosphorus stimulates root growth (there is no shoot growth, ergo it must be stimulating root growth). The unfortunate result of these assumptions is the mantra "Nitrogen for shoots and phosphorus for roots." While there are no nitrogen toxicity symptoms per se, the same cannot be said for phosphate toxicity.

The result of phosphate overfertilizing is leaf chlorosis. Phosphorus is known to compete with iron and manganese uptake by roots, and deficiencies of these two metal micronutrients cause interveinal yellowing. It's my belief that many of the chlorotic shrubs we see in urban landscapes are suffering an indirect iron (or manganese) deficiency from overapplication of phosphorus. Moreover, it has been experimentally demonstrated that high levels of phosphorus are detrimental to mycorrhizal health and lower the rate of mycorrhizal establishment within root systems. This mutually beneficial relationship between the fungus and the plant roots allows the plant to more effectively explore the soil environment and extract needed nutrients. In the absence of mycorrhizae, the plant must expend more energy growing additional roots and root hairs to accomplish the same task.

In addition to harming beneficial soil organisms, excess phosphate will eventually find its way into waterways. Unlike plants in urban landscapes, aquatic plants are most often limited by phosphate, and the addition of phosphate will induce algal blooms (eutrophication). Such blooms are always followed by increased bacterial activity, resulting in lowered oxygen levels and the eventual death of fish and other animals. As green gardeners, it is incumbent upon us to recognize that the excessive use of phosphorus in landscapes is a resource-wasting, ecosystem-damaging practice.

The Bottom Line

- Maintain organic material (mulch) on landscapes; this provides a slow release of phosphorus and other needed macro- and micronutrients over time.
- Don't use phosphate fertilizer when transplanting; in most cases a nitrogen fertilizer is adequate.
- If you have a nutrient deficiency that is not relieved by the addition of nitrogen, try a foliar application of likely nutrients and see if the symptoms are alleviated. While this diagnostic application will only have a temporary effect, it will prevent the excessive addition of mineral nutrients to the soil.

References

Alexander, C., I. J. Alexander, and G. Hadley. 1984. "Phosphate uptake by *Goodyera repens* in relation to mycorrhizal infection." *New Phytologist* 97(3): 401–11.

Biermann, B. J., and R. G. Linderman. 1983. "Effect of container plant growth medium and fertilizer phosphorus on establishment and host growth response to vesicular-arbuscular mycorrhizae." *Journal of the American Society for Horticultural Science* 108(6): 962–71.

Chalker-Scott, L. 2002. "Do anthocyanins function as osmoregulators in leaf tissues?" *Advances in Botanical Research* 37:103–27.

Chalker-Scott, L. 1999. "Environmental significance of anthocyanins in plant stress responses." *Photochemistry and Photobiology* 70:1–9.

Cumbus, I. P., D. J. Hornsey, and L. W. Robinson. 1977. "The influence of phosphorus, zinc and manganese on absorption and translocation of iron in watercress." *Plant and Soil* 48(3): 651–60.

Ducic, T., and A. Polle. 2007. "Manganese toxicity in two varieties of Douglas fir (*Pseudotsuga menziesii* var. *viridis* and *glauca*) seedlings as affected by phosphorus supply." *Functional Plant Biology* 34(1): 31–40.

Hale, V. Q., and A. Wallace. 1960. "Bicarbonate and phosphorus effects on uptake and distribution in soybeans of iron chelated with ethylenediamine di-o-hydroxyphenyl acetate." *Soil Science* 89:285–87.

Johnson, C. R. 1984. "Phosphorus nutrition on mycorrhizal colonization, photosynthesis, growth and nutrient composition of *Citrus aurantium.*" *Plant and Soil* 80(1): 35–42.

Kuo, S., and D. S. Mikkelsen. 1981. "Effect of P and Mn on growth response and uptake of Fe, Mn and P by sorghum." *Plant and Soil* 62(1): 15–22.

Nair, K. P. P., and G. R. Babu. 1973. "Zinc-phosphorus-iron interaction studies in maize." *Plant and Soil* 42(3): 517–36.

Olmsted, S. 2004. "Foliar chlorosis in *Rhododendron* due to excess soil phosphorus." Master's thesis, University of Washington.

Plenchette, C., V. Furlan, and J. A. Fortin. 1983. "Responses of endomycorrhizal plants grown in a calcined montmorillonite clay to different levels of soluble phosphorus. I. Effect on growth and mycorrhizal development." *Canadian Journal of Botany* 61(5): 1377–83.

Topa, M. A., and J. M. Cheeseman. 1993. "32P uptake and transport to shoots in *Pinus serotina* seedlings under aerobic and hypoxic growth conditions." *Physiologia Plantarum* 87(2): 125–33.

Original article posted in September 2000.

THE MYTH OF PHOSPHATE FERTILIZER, PART 2

The Myth

"Roses need phosphate fertilizer for root and flower growth."

I spoke recently at the Northwest Flower and Garden Show on healthy landscape practices. Among my recommendations was the avoidance of phosphate fertilizers, especially at transplant time. After my talk, I was asked if this advice applied to roses too. Not being an expert on roses, my less-than-satisfactory answer was "nonagricultural soils aren't usually deficient in phosphate." The question continued to bother me, however, so I searched the popular and scientific literature for the rose-phosphate connection.

In the popular literature, recommendations are generally like this one: "Dump in a cup of phosphate fertilizer (bonemeal, rock phosphate, superphosphate 0–15–0, or triple phosphate

0–45–0) when planting roses." Books and Web sites alike state that phosphate is required for root establishment and flower production in roses; unfortunately, a great many of these are .edu sites. The number of chemical additives recommended for growing roses in the home landscape is simply astonishing.

The Reality

I was very interested to find no scientific evidence suggesting that roses need high levels of phosphate for any reason. Perennial landscape plants in urban areas are rarely deficient in any nutrient other than nitrogen. In our landscape restoration projects at the University of Washington, soil tests have routinely shown that phosphate levels are at least adequate and sometimes more than adequate for normal plant growth. The addition of any non-deficient nutrient to a landscape is a waste of time and money, and can injure soil organisms, particularly mycorrhizal fungi.

Numerous studies have demonstrated that roses, like most terrestrial plants, maintain a symbiotic relationships with beneficial fungi. If you add phosphate to your roses, you will decrease the ability of mycorrhizal fungi to colonize the roots. Without these fungal partners, the roots must work harder to extract water and nutrients from the soil. Moreover, this excess phosphate is injurious to other soil organisms. With increased fertilizer additions, soil salinity increases. You have then created an artificial system in which soil health is so impaired that you must continue to add fertilizer for your plants to survive.

I believe that this is what has happened in many landscapes featuring roses. Well-intentioned, yet misguided, home owners overapply phosphate and other fertilizers, insecticides, and fungicides until the soil system is so impaired that it becomes nonfunctional. Without the beneficial soil organisms, roses become more susceptible to nutrient deficiencies and opportunistic diseases, causing rose aficionados to add even more of these chemicals.

The Bottom Line

- Nonagricultural soils generally have adequate amounts of phosphate.
- The addition of excess phosphate fertilizer decreases mycorrhizal activity and overall soil health.
- A healthy soil system is better able to support healthy rose growth than one that has been impaired by the overuse of fertilizers and pesticides.

References

Auge, R. M. 1989. "Do VA mycorrhizae enhance transpiration by affecting host phosphorus content?" *Journal of Plant Nutrition* 12(6): 743–53.

Auge, R. M., K. A. Schekel, and R. L. Wample. 1986. "Greater leaf conductance of well-watered VA mycorrhizal rose plants is not related to phosphorus nutrition." *New Phytologist* 103(1): 107–16.

Auge, R. M., and A. J. W. Stodola. 1990. "An apparent increase in symplastic water contributes to greater turgor in mycorrhizal roots of droughted *Rosa* plants." *New Phytologist* 115(2): 285–95.

Davies, F. T. Jr. 1987. "Effects of VA-mycorrhizal fungi on growth and nutrient uptake of cuttings of *Rosa multiflora* in two container media with three levels of fertilizer application." *Plant and Soil* 104(1): 31–35.

Hole, U. B., and G. N. Salunkhe. 1998. "Evaluation of rose cultivars against red scale (*Aonidiella aurantii* Maskell)." *Journal of Maharashtra Agricultural Universities* 22(2): 199–201.

Linderman, R. G., and E. A. Davis. 2004. "Evaluation of commercial inorganic and organic fertilizer effects on arbuscular mycorrhizae formed by *Glomus intraradices*." *HortTechnology* 14(2): 196–202.

Young, T. W., G. H. Snyder, F. G. Martin, and N. C. Hayslip. 1976. "Rose response to nitrogen, phosphorus, and potassium fertilization rates." *Florida Agricultural Experiment Station Technical Bulletin* 771:41.

Young, T. W., G. H. Snyder, F. G. Martin, and N. C. Hayslip. 1973. "Effects of nitrogen, phosphorus, and potassium fertilization of roses on Oldsmar fine sand." *Journal of the American Society for Horticultural Science* 98(1): 109–12.

Original article posted in March 2002.

THE MYTH OF BENEFICIAL BONEMEAL

The Myth

"Add a handful of bonemeal to planting holes before installing shrubs and trees."

Of all the soil amendments on the market, bonemeal seems to be everyone's darling. Credited with stimulating root production and improving flowering, thousands of Web sites promote the use of bonemeal during transplanting and as a regular fertilizer throughout the year. We are assured that bonemeal is "one of the indispensable soil amendments all gardeners should have on hand" and that usage of bonemeal is "good for reducing transplant shock and promoting extensive and healthy root systems." Bonemeal, as the name suggests, is made from animal bones and is favored by organic gardeners and landscapers as a natural source of calcium and phosphorus. There are nearly 10,000 commercial Web sites advertising various formulations of bonemeal. Consumers naturally focus on which of these many products to choose, rather than wonder if any of them are really necessary.

The Reality

Bonemeal is primarily calcium and phosphorus, two elements which are usually adequate in nonagricultural soils. The N-P-K (nitrogen-phosphorus-potassium) analyses of bonemeal preparations vary, but are generally in the range of 0–12–0 to 3–20–0. Both calcium and phosphorus are required for plant growth, but both (and especially phosphorus) can cause problems if they occur in high concentrations. It is important to understand that neither element, nor any other mineral, will "stimulate" plant growth beyond what is normal for a particular plant.

Why does the myth of phosphorus-induced root stimulation persist? The answer probably lies in the effect phosphorus fertilizers have on mycorrhizal relationships. When plant roots are in low-phosphorus environments, they exude organic acids from their root tips. These acids allow mycorrhizal fungi to penetrate the roots and form the networks that assist plant roots in taking up water and nutrients. Mycorrhizae are particularly adept at extracting phosphorus from the soil.

If phosphorus levels are too high, however, the roots do not exude the organic acids, and mycorrhizal connections do not form. This forces the plant to put more resources into root growth to compensate for the lack of mycorrhizae. So, in a sense, phosphorus will increase root growth—but at an added cost to the plant. The resources expended by the plant in growing additional roots to take the place of mycorrhizae are not available for other plant needs.

Shrub and tree species that are mycorrhizae-dependent have a difficult time surviving in soils where mycorrhizae cannot develop. In particular, seedlings and newly transplanted materials are less efficient in absorbing water and minerals from the soil and are more likely to suffer transplant shock than plants where mycorrhizae are present. Adding mycorrhizal spores to soils where phosphorus is too high is ineffective—the spores will remain dormant. Interestingly, bonemeal (and other phosphorus sources) is toxic to members of the Protea family. These plants and others have adapted to nutrient-poor soils and can easily scavenge necessary minerals. This natural ability is compromised when fertilizers are overapplied.

What can you do if you have added too much phosphorus over the years? If a soil test indicates that phosphorus levels are high, you may be able to tie up excess phosphorus by adding a mixture of other mineral fertilizers. I've not had to do this myself, but various Web sites recommend concoctions of ammonium sulfate, magnesium sulfate (Epsom salts), iron sulfate, and zinc sulfate. Caution should be exercised before adding any minerals, as further nutrient imbalances may occur; it would be best to ask for advice from a soil scientist. In any case, levels of soil phosphorus will eventually decrease if phosphorus-containing fertilizers are discontinued.

The Bottom Line

- Bonemeal supplies high levels of phosphorus and calcium, elements that are rarely limiting in nonagricultural soils.

- Phosphorus, from bonemeal or other sources, does not "stimulate" plant growth; it is only a mineral, not a plant growth regulator.

- High levels of phosphorus, from bonemeal or other sources, will inhibit growth of mycorrhizal fungi.

- Without mycorrhizal partners, plants must put additional resources into root growth at the expense of other tissues and functions.

- Before you add any supplementary nutrients to your landscape, have a complete soil test performed first.

References

Criley, R. A. 2001. "Proteaceae: Beyond the big three." *Acta Horticulturae* 545:79–85.

Montarone, M., and M. Ziegler. 1997. "Water and mineral absorption for two protea species (*P. eximia* and *P. cynaroides*) according to their development stage." *Acta Horticulturae* 453:135–44.

See also references in "The Myth of Phosphate Fertilizer, Part 1" and "The Myth of Phosphate Fertilizer, Part 2."

Original article posted in May 2004.

THE MYTH OF POLYACRYLAMIDE HYDROGELS

The Myth

"Polyacrylamide hydrogels are environmentally safe substances that reduce irrigation needs."

With the possibility of summer drought always looming on the horizon, especially given global warming, those of us whose business or pleasure includes landscape plants are understandably concerned about water issues. In response, .com Web sites are full of products promising to reduce water usage in the landscape. Prominent among these products are hydrogels, which have been used successfully by the landscape industry to reduce transplant shock and increase containerized plant growth. These hydrogels, sometimes referred to as "root-watering crystals" or "water-retention granules," swell like sponges to several times their original size when hydrated. Water is then released slowly to the surrounding soil, reducing the need for irrigation.

Once considered to be a professional nursery product, hydrogels are increasingly popular with home owners, who add them to vegetable gardens, container plants, annual beds, lawns, and perennial landscapes. The most commonly available hydrogels are polymers of acrylamide and potassium acrylate. These polymers have a longer functional life (perhaps up to five years) than other organic hydrogels composed of starch, gelatin, or agar.

The Reality

My initial concern with hydrogel usage was with the public perception that it is a permanent fix. But after five years, virtually all hydrogel will be depolymerized through natural decomposition processes, and, like a sponge cut into miniscule pieces, will no longer absorb water. The rate of degradation is increased in the presence of fertilizer salts (and no, it doesn't make any difference if these fertilizers are synthetic or organic). One is then left with the original soil conditions. In a permanent landscape, this can be problematic unless other water-conserving steps are then implemented.

But my second, and greater, concern arose when I discovered that hydrogels are constructed of acrylamide units. Hydrogels are routinely touted as pH-neutral, nontoxic, environmentally friendly compounds, which they are in their polymerized form. But when hydrogels break down, they depolymerize, releasing potassium acrylate and acrylamide. Though it is quickly degraded in the environment, acrylamide is a lethal neurotoxin and has been found to cause cancer in laboratory animals. It readily passes through the skin and can be inhaled as dust. Unfortunately, the chemical data sheets on hydrogels do not mention the fact that after only a few years they will be completely decomposed, forming acrylamide units and other unknown products along the way. Since polyacrylamide is defined as "not readily biodegradable" (less than 10 percent is degraded after twenty-eight days), some sellers of hydrogels actually promote their products as "nonbiodegradable"!

Who is at risk of acrylamide exposure? Workers in the nursery and landscape industry who routinely use hydrogels may become exposed to them as they degrade and become toxic. Home owners who add hydrogel-containing potting mix to their landscapes or compost piles are exposed. Dogs, cats, and wildlife that come in contact with these substances are at risk. On a larger scale, entire ecosystems are at risk as acrylamide and smaller degradation products are water-soluble and can easily enter watersheds.

One of the greatest pleasures of gardening is getting your hands into good, rich soil and breathing in its aroma. I believe that the increased, and indiscriminate, use of polyacrylamide hydrogels is an extremely serious hazard to human health and to the environment.

The Bottom Line

- Hydrogels are organic compounds that will degrade after two to five years; they are not a long-lasting solution to droughty conditions.

- Exposure to fertilizer salts will increase the degradation rate of hydrogels.

- When hydrogels degrade, one of the byproducts is acrylamide, a deadly neurotoxin and potential carcinogen.

- Acrylamide can be absorbed through the skin or by inhaling; people who have a likely risk of exposure to this compound require safety clothing and dust masks.

- There are safe (albeit shorter-lived) alternatives to polyacrylamide hydrogels, including starch-based gels and other gels currently used in cosmetic surgery.

- There are better, more environmentally sound ways to reduce water usage and improve water retention of soils than through hydrogels, including careful plant selection and mulching.

References

As of this writing, there are hundreds of papers covering various aspects of polyacrylamide gels used in the landscape. For the sake of brevity, I've included a very limited number of the most recent and/or relevant publications on this topic.

Environmental Degradation of Polyacrylamide Gels

Grula, M. M., M. L. Huang, and G. Sewell. 1994. "Interactions of certain polyacrylamides with soil bacteria." *Soil Science* 158(4): 291–300.

Holliman, P. J., J. A. Clark, J. C. Williamson, and D. L. Jones. 2005. "Model and field studies of the degradation of cross-linked polyacrylamide gels used during the revegetation of slate waste." *Science of the Total Environment* 336(1–3): 13–24.

Kay-Shoemake, J. L., M. E. Watwood, R. D. Lentz, and R. E. Sojka. 1998. "Polyacrylamide as an organic nitrogen source for soil microorganisms with potential effects on inorganic soil nitrogen in agricultural soil." *Soil Biology and Biochemistry* 30 (8–9): 1045–52.

Mai, C., W. Schormann, A. Majcherczyk, and A. Huttermann. 2004. "Degradation of acrylic copolymers by white-rot fungi." *Applied Microbiology and Biotechnology* 65:479–87.

Matsuoka, H., F. Ishimura, T. Takeda, and M. Hikuma. 2002. "Isolation of polyacrylamide-degrading microorganisms from soil." *Biotechnology and Bioprocess Engineering* 7(5): 327–30.

Smith, E. A., S. L. Prues, and F. W. Oehme. 1996. "Environmental degradation of polyacrylamides. I. Effects of artificial environmental conditions: Temperature, light, and pH." *Ecotoxicology and Environmental Safety* 35:121–35.

Smith, E. A., S. L. Prues, and F. W. Oehme. 1997. "Environmental degradation of polyacrylamides. II. Effects of environmental (outdoor) exposure." *Ecotoxicology and Environmental Safety* 37(1): 76–91.

Stahl, J. D., M. D. Cameron, J. Haselbach, and S. D. Aust. 2000. "Biodegradation of superabsorbent polymers in soil." *Environmental Science and Pollution Research* 7(2): 83–88.

Taban, M., and S. A. R. M. Naeini. 2006. "Effect of Aquasorb and organic compost amendments on soil water retention and evaporation with different evaporation potentials and soil textures." *Communications in Soil Science and Plant Analysis* 37(13–14): 2031–55.

Failure of Polyacrylamide Gels in Hydrating Plants

Abbey, T., and T. Rathier. 2005. "Effects of mycorrhizal fungi, biostimulants and water absorbing polymers on the growth and survival of four landscape plant species." *Journal of Environmental Horticulture* 23(2): 108–11.

Booth, D. T. 2005. "Establishing Wyoming big sagebrush seed orchards on reclaimed mined land." *Native Plants Journal* 6(3): 247–53.

Gilman, E. G. 2004. "Effects of amendments, soil additives, and irrigation on tree survival and growth." *Journal of Arboriculture* 30(5): 301–10.

Rowe, E. C., J. C. Williamson, D. L. Jones, P. Holliman, and J. R. Healey. 2005. "Initial tree establishment on blocky quarry waste ameliorated with hydrogel or slate processing fines." *Journal of Environmental Quality* 34(3): 994–1003.

Stamps, R. H., and D. M. McColley. 2005. "Effects of soil incorporation of polyacrylamide copolymer on newly planted leatherleaf fern (*Rumohra adiantiformis* [Forst.] Ching)." *Proceedings of the Interamerican Society for Tropical Horticulture* 48:173–76.

Wallace, A., G. A. Wallace, and A. M. Abouzamzam. 1986. "Effects of excess levels of a polymer as a soil conditioner on yields and mineral nutrition of plants." *Soil Science* 141(5): 377–80.

Polyacrylamide Gel Toxicity Concerns

Beim, A. A., and A. M. Beim. 1994. "Comparative ecological-toxicological data on determination of maximum permissible concentrations (MPC) for several flocculants." *Environmental Technology* 15(2): 195–98.

Fraire, A. E., I. Shahab, S. D. Greenberg, A. Jubran, and M. Noal. 1992. "Experimental polyacrylamide-induced acute injury in rat lung." *Chest* 102(5): 1591–94.

Hasegawa, R., K. Naitoh, Y. Kawasaki, Y. Aida, J. Momma, M. Saitoh, Y. Nakaji, and Y. Kurikawa. 1990. "Acute and subacute toxicity studies on 2,3–dichloropropionic acid and chlorinate polyacrylamide in rats." *Water Research* 24(5): 661–66.

King, D. J., and R. R. Noss. 1989. "Toxicity of polyacrylamide and acrylamide monomer." *Reviews on Environmental Health* 8(1–4): 3–16.

Schneider, R. J., R. L. Lightle, R. G. Reddy, A. Rehemtulla, B. D. Ross, R. Kopelman, and M. A. Philbert. 2003. "Acute toxicity of polyacrylamide and sol-gel nanoparticles in rats." *Toxicological Sciences* 72(S-1): 304.

Original article posted in June 2001.

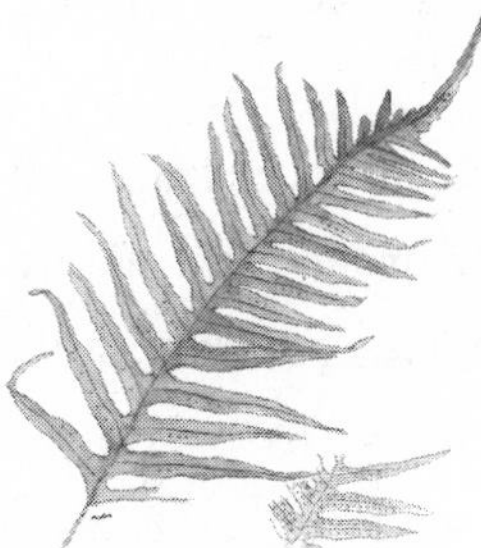

THE MYTH OF POLYACRYLAMIDE HYDROGELS REVISITED

The Myth

"Are polyacrylamide hydrogels environmentally safe substances that reduce irrigation needs?"

When my column first ran in June 2001, I was threatened with a lawsuit from a hydrogel manufacturer. Fortunately, I was able to substantiate my comments with journal literature, and the issue was dropped. Since then, I continue to receive e-mails from students, Master Gardeners, nursery and landscape professionals, medical doctors, researchers, and concerned individuals all over the world. It's time to look at the recent literature and see what's new.

Polyacrylamide (PAM) hydrogels help create larger soil aggregates that reduce erosion and improve water-holding capacity. Water is released slowly to the surrounding soil, reducing the need for irrigation. In addition to their soil applications, PAM hydrogels are also used extensively in municipal water treatment, paper manufacturing, pesticide formulations, and food processing.

The Reality

Briefly, here is what is currently known about polyacrylamide hydrogels (PAM):

Characteristics

- PAM hydrogels are made of acrylamide, a potent carcinogen and neurotoxin. During manufacture, there is a small percentage of acrylamide still present (generally 0.01–0.05 percent), meaning that new hydrogels will always contain a small portion of acrylamide.

- PAM hydrogels are not considered hazardous (but see below), but several of the identified degradation byproducts are.

Environmental Interactions

- PAM hydrogels do not work well in clay soils.

- Neither PAM, acrylamide, nor any other degradation product appears to be taken up by plant roots.

- Fertilizer salts and saline soils decrease the water uptake and holding capacity of PAM hydrogels.

- Positively charged ions (cations) such as potassium, calcium, magnesium, and iron decrease water absorption by PAM hydrogels by as much as 90 percent. These minerals are plant nutrients that naturally occur in the soil and are contained in fertilizers.

- Since lower (more acidic) pH levels increase the solubility of cations (especially metals such as magnesium and iron), hydrogels in acidic soils are even more likely to be ineffective.

- PAM hydrogels were found to decrease plant uptake of several essential nutrients in field studies.

Degradation

- Both fungal and bacterial species commonly found in the soil are capable of degrading PAM hydrogels. In the lab this degradation can be rapid, but is probably slower in the field.

- Ultraviolet radiation from sunlight will degrade PAM hydrogels.

- Tilling and other shearing forces will degrade PAM hydrogels.

- Acrylamide can be a degradation product, though it is detoxified in a matter of days through the degradation process.

- Though PAM hydrogel degradation may not generate large amounts of acrylamide, it can produce acrylonitrile (an EPA-regulated substance) and unknown, uncharacterized polyacrylate units of various sizes whose health and environmental effects are likewise unknown.

- Lack of information on the possible toxicity of degraded PAM hydrogels makes it impossible to assess health risks to humans, animals, or ecosystems.

Impacts on Aquatic and Human Health

- PAM hydrogels used as flocculants in water treatment systems affect all water ecosystem components, especially certain types of algae, invertebrates, and adult fish.

- Contamination with acrylamide prevents PAM hydrogels from being used in drinking water treatment or for medical applications (exposing a troubling difference in safety standards).

- Even though they are considered nontoxic, PAM hydrogels can cause lung injury if inhaled.

- A plastic surgeon in Taiwan informed me that PAM hydrogels are routinely used for human tissue enhancement in China, Eastern Europe, and the Balkans. He has seen firsthand evidence of the damage done when PAM hydrogels are injected into the human body. (It speaks volumes that in this country we don't use PAM gels in plastic surgery—we use starch-based compounds.)

From the available scientific literature, it appears that PAM hydrogels are not as ubiquitously useful for field applications as was previously assumed. There are a number of environmental factors that hamper the functionality of PAM hydrogels, and their environmental degradation is firmly substantiated. What is actually produced during this degradation is unknown, as is the environmental and human-health impact.

There are alternatives to PAM hydrogels that could be adapted to landscape uses. Recently, a research group explored using waxy corn and tapioca starches instead of PAM hydrogels for oil spill recovery. They found these starches to be more efficacious, more economical, and more environmentally sound. Here's hoping that the next time this subject is reviewed, some of these alternatives will have become industry standards.

The Bottom Line

- PAM hydrogels are widely used in a number applications through which humans, animals, and ecosystems may be exposed.
- Many environmental factors can limit the efficiency of PAM hydrogels and speed their degradation.
- PAM hydrogels are not long-term solutions to droughty conditions.
- While the toxicity of acrylamide is well known, the hazards posed by PAM hydrogel degradation products are not.
- There are agricultural alternatives to PAM hydrogels, including starch-based gels.

References

See references in "The Myth of Polyacrylamide Hydrogels."

Original article posted in March 2004.

THE MYTH OF WANDERING WEEDKILLER

The Myth

"Glyphosate will move through root grafts and kill nontarget plants."

An increasingly common concern in landscape restoration is inadvertent exposure of desirable, nontarget trees and shrubs to herbicides. To reduce injury to nontarget species, pesticide applicators often avoid sprays and instead paint the freshly cut ends of stumps and canes of weedy species with glyphosate (the active ingredient in Roundup). The herbicide translocates through the exposed phloem elements in the stem to the roots, ideally killing the target weed and leaving neighboring vegetation untouched.

A warning on the product sheet for Roundup reads, "Avoid painting cut stumps with this product as injury resulting from root grafting may occur in adjacent trees." This warning has given rise to concerns that translocatable herbicides like glyphosate may move underground between the roots of different plants (root grafts) and kill nontarget vegetation. Various Web sites make similar statements: "Instances have been documented where a herbicide has moved from a treated tree to another of the same species or genera through a root graft." If these claims are true, how can we continue to manage aggressive, invasive weeds like blackberry (*Rubus discolor*) and Japanese knotweed (*Polygonum cuspidatum*)?

The Reality

Though the aboveground portions of plants tend to grow in isolation as they compete for sunlight, the landscape below ground is radically different. Roots from the same, and sometimes different, species fuse as they contact one another. Threads of symbiotic fungal hyphae (mycorrhizae) increase the connectedness of this underground community even further. The result is an intricate network of plant roots and their fungal partners, among which water, minerals, and carbohydrates may freely move. This fascinating underground ecosystem appears to be responsible for the documented survival of completely girdled trees, which have stayed alive as a result of carbohydrate transport from intact neighbors.

Do herbicides move through these networks? There are only a few studies that have addressed this question, but fortunately each relied on evidence from hundreds or thousands of treated stems or trees. The first study was conducted several decades ago using sodium arsenate to prevent regrowth of aspen (*Populus*) and various oak (*Quercus*) species. Several hundred adjacent trees were studied for damage from translocation of the poison from the roots of the target tree to others in the root network, but no symptoms of herbicide poisoning were observed. More recently, researchers applied glyphosate, imazapyr, or triclopyr to control regrowth of cut stumps of ash (*Fraxinus*), sycamore (*Platanus*), and birch (*Betula*). Again, no evidence was found to suggest root translocation of any of the herbicides—applied to thousands of cut stems—to nearby untreated trees.

On the other hand, disease vectors (organisms that transmit pathogens, most notably the fungi responsible for Dutch elm disease, oak wilt, and other fungal diseases) do cross root grafts. Several independent studies have documented the transmission of a disease from infected trees through root networks to adjacent, healthy relatives. In such cases severing roots, or using chemical or physical root barriers, can halt fungal spread.

Why do disease vectors move through root grafts, but herbicides don't? Pathogens, like the fungi mentioned earlier, require new hosts for their survival. As living organisms, they can evolve biochemical pathways that allow them to overcome natural plant barriers and attack fresh hosts. Herbicides, on the other hand, are not living organisms and do not change their chemical structure. Though we understand little about how substances translocate through grafted roots, it would not be surprising to find sophisticated barriers within grafted roots to monitor the passage of substances from plant to plant. It is also probably true that healthy trees are more likely to have intact barriers to chemical translocation, while those already impacted by fungal disease may not. Herbicides might be able to translocate through grafted roots already breached by fungal degradation; perhaps this is why the warning exists on Roundup.

Finally, it's important to realize that root grafts tend to be most common between trees of the same species, or occasionally within the same genus. It would be highly unlikely to find the roots of *Rubus discolor* or other weed species grafted to the roots of unrelated trees or shrubs. Therefore, glyphosate applied to the cut canes of blackberry or other common weedy shrubs would not be expected to move into neighboring vegetation.

The Bottom Line

- Field research indicates that glyphosate and other translocatable herbicides do not cross root grafts in healthy trees.
- Fungal vectors can breach root grafts through degradative enzymatic activity.
- Root grafts that have already been breached by fungi may serve as conduits for herbicide translocation as well.
- Unrelated plants are unlikely to form root grafts.

References

Dickinson, G. R., and J. R. Huth. 2003. "Susceptibility of *Corymbia citriodora* subsp. *variegata, Eucalyptus cloeziana* and *E. grandis* to the cut-stump application of glyphosate following pre-commercial thinning operations in North-Eastern Australia." *Journal of Tropical Forest Science* 15(3): 505–09.

Dixon, F. L., and D. V. Clay. 2002. "Imazapyr application to *Rho-*

dodendron ponticum: Speed of action and effects on other vegetation." *Forestry* 75(3): 217–25.

Willoughby, I. 1999. "Control of coppice regrowth in roadside woodlands." *Forestry* 72(4): 305–12.

Original article posted in November 2002.

MULCHES

THE MYTH OF LANDSCAPE FABRIC

The Myth

"Landscape fabric provides permanent weed control for ornamental landscapes."

Increased concern over the indiscriminate use of herbicides has caused landscape professionals and consumers to look closely at nonchemical alternatives to weed control. Mulches are increasing in popularity as weed-control strategies and have a number of additional benefits, including water retention and soil protection. Mulches may be organic, inorganic, or synthetic and can often bring an aesthetic quality in tandem with their principal role in plant-health maintenance. Synthetic mulches, including geotextiles, are of interest to many consumers and professionals because they are perceived as nonbiodegradable, permanent solutions to weed control.

Initially developed for agricultural use, geotextiles have found their way into ornamental installations as landscape fabrics. These fabrics, a vast improvement over the impermeable black plastics still (unfortunately) used for weed control, are woven in such a way that water and gas exchange can occur but light penetration is significantly reduced. Hence, they are effective in reducing weed seed germination in areas where soil disturbance would otherwise induce the germination of a horde of weeds. Such fabrics have been so effective in reducing weeds in vegetable and ornamental crop production that they have been applied to more permanent landscape installations.

The Reality

Like the perpetual dieter searching for a permanent weight-loss pill, we as gardeners continue to seek permanent weed-control solutions. Unfortunately, there is no such permanent fix. We must remain "ever vigilant" in our battle with weeds and cannot rely on a product to do this passively. The fact is that weed-control fabrics are not permanent and will decompose, especially when exposed to sunlight. Such fabrics are effective in agricultural situations, in annual planting beds, or where the landscape is regularly disturbed and the fabrics can be replaced when needed. For permanent landscapes, however, they are not a long-term solution and, in fact, can hinder landscape plant health. Some of the documented drawbacks of these fabrics are listed below:

- Geotextiles degrade in the landscape in as little as one year if unprotected from sunlight. The tattered remnants are unattractive and, when removed, can damage surrounding plants.

- Any organic matter or soil on top of the fabrics will quickly be colonized by weeds; this precludes covering the fabric with anything but inorganic mulch like pebbles. It also requires continual maintenance to keep the fabric free of debris.

- Weeds will eventually grow on top of and through these fabrics, making their removal difficult.

- Landscape plant roots can also colonize fabrics, and they are damaged when the fabrics are removed.

- The aesthetic quality of landscape fabrics is minimal; it becomes worse as the materials begin to degrade.

I must add my own anecdotal story here. When we moved into our current house a few years ago, I began attacking the horsetail and bindweed that were emerging from our backyard ornamental bed. The odd thing was my shovel wouldn't go deeper than about six inches. I discovered that the previous owners had laid landscape fabric down and then covered it with six inches of topsoil. The fabric was completely covered with the roots and rhizomes of both bindweed and horsetail. It was apparent that the owners had hoped to cover up the problem long enough to sell the house (I guess it worked!). Up came the fabric, out came the roots and rhizomes, and down went the wood chips. Now, a year later, the bed is nearly 100 percent free of both of these weeds, thanks to the wood chips and being "ever vigilant."

The Bottom Line

- Geotextiles are not effective weed-control solutions for permanent landscapes.
- Landscape fabrics used in permanent landscape installations will eventually become high maintenance problems in terms of appearance, weed control, and plant health.
- Organic mulches are preferable for permanent landscape installations, as they can be reapplied throughout the life of the landscape without damaging the existing plants.

References

Appleton, B. L., J. F Derr, and B. B. Ross. 1990. "The effect of various landscape weed control measures on soil moisture and temperature, and tree root growth." *Journal of Arboriculture* 16:264–68.

Horowitz, M. 1993. "Soil cover for weed management." *Communications of the 4th International Conference I.F.O.A.M., Non-chemical Weed Control,* ed. J. M. Thomas. Association Colloque IFOAM, Dijon, France, 149–54.

Houle, G., and P. Babeux. 1994. "Fertilizing and mulching influence on the performance of four native woody species suitable for revegetation in subarctic Quebec." *Canadian Journal of Forest Research* 24:2342–49.

Original article posted in January 2002.

USING ARBORIST WOOD CHIPS FOR WEED CONTROL

- If there are concerns about disease, let wood chips age before using them.
- If there are concerns about nutrient deficiencies, create a thin underlying layer of a more nutrient-rich mulch (like compost) before installing wood chips.
- Begin mulch application before annual weeds are established (in spring or fall).
- Remove perennial weeds in early spring, when root resources are lowest.
- Prune or mow perennial weeds at the root crown; pulling destroys soil structure.
- Remove all noxious weed materials from the site to prevent rerooting.
- Thick layers (4–6" for ornamental sites, 8–12" for restoration sites and to control aggressive perennial weeds such as blackberry and ivy) of coarse materials are best for weed control and water conservation.
- Keep mulch away from the trunks of trees and shrubs.
- Pull any resprouting weeds; the mulch layer prevents erosion and facilitates pulling, as compared to unmulched soil.
- Replace mulch as needed to maintain the appropriate depth (a minimum depth of 8 inches for weed control in low-maintenance sites).
- The replacement rate depends on the decomposition rate.

THE MYTH OF CLEAN COMPOST

The Myth

"Compost is a safe, chemical-free source of nutrients for gardens."

Woe to anyone who does not embrace the wonders of compost! Commercial compost is advertised as "environmentally sound," "all natural," "chemical-free," "fish friendly," et cetera. Glossy photographs demonstrate the results we can expect if we use compost in our flower beds, vegetable gardens, and perennial landscapes. For those of us who prefer to make our own, there are many how-to guides to lead us past the pitfalls that can cause our compost to become contaminated with weed seed or pathogens. Whether we make our own or purchase it commercially, we are led to believe that compost is safe for our soils, our plants, and ourselves.

The Reality

Compost is great stuff. (Of course it is not chemical free; all substances are made of chemicals.) Under ideal conditions, it is a safe, environmentally friendly way of recycling yard waste and returning nutrients to the soil. The telltale sharp, sour odor helps identify a "bad" batch of compost. There are other contaminants, however, whose presence is unexpected, subtle, and injurious to plant and human health.

I am particularly concerned with two classes of compost contaminants: pesticides and heavy metals. Recently, compost feedstocks in eastern Washington were found to be contaminated with clopyralid and picloram—two broadleaf herbicides. These relatively persistent herbicides have been found in hay and grain residues and the manure of chickens, cattle, and horses. Compost contaminated with these herbicides can injure or kill broadleaf ornamentals and vegetables. While the source of this particular contamination problem was agricultural, these broadleaf weed killers are also used for lawn care. If treated lawn clippings are composted, either at home or elsewhere, they will contaminate the compost and kill or injure susceptible plants. The long-term impacts of these herbicides on human health are not yet known.

Heavy metals, such as lead, arsenic, and mercury, are less problematic for plants than they are for humans. If ingested, these metals disrupt biochemical pathways and cause a number of health problems, particularly in children. Lead is the most commonly found heavy metal in residential urban soils, primarily as a remnant of lead-based paints and fuels. In a recent worldwide study of compost, frighteningly high levels of lead, chromium, cadmium, manganese, and other EPA-regulated heavy metals were reported. Sources of toxic heavy metals include sewage sludge, industrially contaminated soil, and the previously mentioned lead problem. It's not a bad idea to have your soil tested for lead, especially if you grow produce for human consumption. The cost is minimal, and the information invaluable.

If you depend on commercially available compost, be aware that the U.S. Composting Council has a Seal of Testing Assurance program. Members in this program must test their products for pesticides, heavy metals, and pathogens on a regular basis. These numbers are available to the public. More information can be found at this Web site: http://tmecc.org/sta/index.html.

The Bottom Line

- The best sources for pesticide-free compost are those that have been analyzed and certified. Homemade compost is also a good choice, as long as you are sure that your materials are contaminant-free.

- Unregulated compost can contain pesticides, heavy metals, and other environmental toxins that may be harmful to you and your plants.

- Soil testing for heavy metals is crucial for any landscape where plants are grown for human consumption.

- If you must have your lawn sprayed with persistent, broadleaf herbicides, be sure to use a mulching mower and leave the clippings in place. Do NOT compost them or bag them for clean green removal.

References

DeMiguel, E., M. J. DeGrado, J. F. Llamas, A. Martin-Dorado, and L. F. Mazadiego. 1988. "The overlooked contribution of compost application to the trace element load in the urban soil of Madrid (Spain)." *Science of the Total Environment* 215(1–2): 113–22.

Downer, J., A. Craigmill, and D. Holstege. 2003. "Toxic potential of oleander derived compost and vegetables grown with oleander soil amendments." *Veterinary and Human Toxicology* 45(4): 219–21.

Fauci, M., D. F. Bezdicek, D. Caldwell, and R. Finch. 2002. "Development of plant bioassay to detect herbicide contamination of compost at or below practical analytical detection limits." *Bulletin of Environmental Contamination and Toxicology* 68(1): 79–85.

Gatto, O., and B. Pavoni. 1999. "Compost amending capability and effects on soil salinity and heavy metal content in an open field cultivation of *Eruca sativa*." *Toxicological and Environmental Chemistry* 73(3–4): 221–35.

Genevini, P. L., F. Adani, D. Borio, and F. Tambone. 1993. "Heavy metal content in selected European commercial composts." *Compost Science and Utilization* 5(4): 31–39.

Kapanen, A., and M. Itavaara. 2001. "Ecotoxicity tests for compost applications." *Ecotoxicology and Environmental Safety* 49(1): 1–16.

Kovacic, D. A., R. A. Cahill, and T. J. Bicki. 1992. "Compost: Brown gold or toxic trouble?" *Environmental Science and Technology* 26(1): 38–41.

Maynard, A. A. 1998. "Utilization of MSW compost in nursery stock production." *Compost Science and Utilization* 6:38–44.

McCartney, D., Y. Zhang, and C. Grant. 2001. "Characterization of compost produced at a golf course: Impact of historic mercury accumulations in putting green soil." *Compost Science and Utilization* 9(1): 73–91.

Strom, P. F. 1998. "Evaluating pesticide residues in yard trimmings compost." *Biocycle* 39(11): 80.

Wagman, N., B. Strandberg, B. van Bavel, P. A. Bergqvist, L. Oberg, and C. Rappe. 1999. "Organochlorine pesticides and polychlorinated biphenyls in household composts and earthworms (*Eisenia foetida*)." *Environmental and Toxicological Chemistry* 18(6): 1157–63.

Original article posted in February 2002.

HOW DOES MULCHING REDUCE PESTICIDE AND FERTILIZER USE?

- Mulches reduce weed seed germination and growth, reducing the need for herbicides. Additionally, lack of weed competition reduces need for fertilizers.
- Mulches enhance populations of beneficial microbes and other soil organisms, which compete with many pathogens and pests, reducing the need for fungicides and insecticides.
- Mulches provide habitat for predatory insects, reducing the need for insecticides.
- Organic mulches provide a natural release of nutrients available to plants, reducing the need for fertilizers.
- Organic mulches provide an environment for beneficial mycorrhizae on the roots that assist in nutrient uptake, reducing the need for fertilizers.

THE MYTH OF PAPER-BASED SHEET MULCH

The Myth

"Newspaper and cardboard sheet mulches are excellent ways to reduce weeds and maintain soil health in permanent landscapes."

In their quest to create more sustainable landscapes (those that require less input of fertilizers, pesticides, and other resources) gardeners, landscapers, and restoration ecologists have focused on mulches. Of particular interest are organic mulches, and even more appealing are those that recycle materials that might otherwise contribute to landfills.

The use of mulches to suppress weeds and conserve soil water has a substantial agricultural history. Newspaper mulch, either as intact sheets or chopped and shredded, has been successful in reducing weeds and increasing yield in some row crops. Cardboard sheet mulch, often used in tree plantations, has been less reliable. These paper mulches are increasingly common in urban landscapes, especially in restoration sites. Are they effective in suppressing weeds, maintaining soil water, and aiding plant establishment in this context?

The Reality

The use of newspaper and cardboard sheet mulches in non-crop settings is relatively new, and therefore not much scientific literature exists on its efficacy in permanent landscapes. However, there are some caveats from the agricultural literature as well as anecdotal observations that can be applied to permanent installations:

- Newspaper and cardboard sheet mulches can become pest havens. Termites were found to prefer cardboard over wood chips as a food source, and rodents such as voles often nest underneath mulch sheets.

- Newspaper and cardboard sheet mulches were often not as effective as other organic mulches (e.g., wood chips or bark) in preventing weed growth or improving yield.

- Newspaper and cardboard sheet mulches often become dislodged by winds, especially if they are exposed.

- Newspaper and cardboard sheet mulches can induce anaerobic conditions if used on wet, poorly drained soils. When wet, the layers of paper are compacted, creating an impermeable barrier to water and gas exchange.

- Newspaper and cardboard sheet mulches become hydrophobic if allowed to dry out, causing rainfall or irrigation water to sheet away rather than percolate through. This is particularly true in regions with droughty summers or well-drained soils.

- Newspaper and cardboard sheet mulches are not aesthetically appealing when exposed.

Newspaper and cardboard sheet mulches have been effectively used in home gardens where soil is continuously worked and irrigation is applied. For less-well-maintained sites, they are not the best choice for the reasons given above. I have observed increased shrub death in restoration areas mulched with newspaper and cardboard sheet mulches, compared to adjacent sites where wood chips have been used. Regions that experience dry summers, especially sites with well-drained soils, are contraindicative for the use of sheet mulches in permanent landscape installations.

The Bottom Line

- Newspaper and cardboard sheet mulches can be effective for annual beds if they are properly maintained.

- Sheet mulches can prevent water movement and gas exchange if they are too wet or too dry.

- Use site-appropriate mulch materials. Permanent, ornamental landscapes, nonmaintained sites, and restoration areas are not appropriate locations for newspaper and cardboard sheet mulches.

References

Buschbeck, T. 1991. "Mulch plates against weed competition in broadleaved plantations." *Allegemeine Forst Zeitschrift* 46(19): 974–75.

Calkins, J. B., B. T. Swanson, and D. L. Newman. 1996. "Weed control strategies for field grown herbaceous perennials." *Journal of Environmental Horticulture* 14(4): 221–27.

Long, C. E., B. L. Thorne, and N. L. Breisch. 2001. "Termites and mulch." *Pest Control Technology* 29(10): 64–70.

Schwarz, M., and M. Mussong. 1991. "Time requirements for putting out cardboard mulch plates." *Allegemeine Forst Zeitschrift* 46(19): 970–72.

Siipilehto, J. 2001. "Effect of weed control with fibre mulches and herbicides on the initial development of spruce, birch and aspen seedlings on abandoned farmland." *Silva Fennica* 35:403–14.

Warmund, M. R., C. J. Starbuck, and C. E. Finn. 1995. "Micropropagated 'Redwing' raspberry plants mulched with recycled newspaper produce greater yields than those grown with black polyethylene." *Journal of Small Fruits and Viticulture* 3(1): 63–73.

Original article posted in June 2002.

THE MYTH OF PRETTY MULCH

The Myth

"Bark mulch and sawdust are aesthetically preferable to wood chips, and they work just as well."

When I speak to professional and community groups, I am invariably asked if bark mulch and sawdust can be used in place of arborist wood chips. Initially this question surprised me, given that bark mulch and sawdust can be quite expensive and that wood chips are cheap, if not free. When I press people on this issue, it turns out that the real reason they want to use bark mulch and sawdust is aesthetic.

People like neat landscapes—no weeds, no bugs, no leaf litter. It's another way for us to separate ourselves from the "messiness" of nature. And most of us realize that leaving soil unprotected is not a good management practice. Therefore, bark mulch and sawdust are viewed as desirable mulches because they are uniform in color and texture—you can even buy colorized sprays to restore weathered mulch to its original appearance! The USDA promotes the use of bark mulch for "attractive" landscapes.

The Reality

The "invention" of bark and sawdust mulch was beneficial to both the landscape and timber industries. Prior to this time, the timber industry used these lumber leftovers as hog fuel. Recycling these materials in a more environmentally friendly way theoretically benefits everyone. There are, however, some problems associated with bark and sawdust mulches that must be recognized by the landscape industry and by home owners.

First of all, bark does not function like wood chips in its water-holding capacity. Bark is the outer covering of the tree, and contains suberin, a waxy substance that prevents water loss from the trunk. Suberin also prevents external water from moving into the bark, and, in fact, helps explain why fresh bark mulch always seems dry. Wood chips, on the other hand, consist primarily of the inner wood, which is not suberized and has the capacity to absorb and hold moisture. One perceptive gardener at the University of Washington realized that bark mulch had created a "nearly impenetrable wall between surface water and plant roots" and replaced the bark with wood chips, which increased water flow through the mulch layer. Due to its fine texture, sawdust also creates an impermeable barrier, which repels rain and irrigation water. Mulches, like soil, benefit from a variety of particle sizes to facilitate water and air movement.

Second, bark mulch is often a source of weed infestation. While newer lumber mills have clean areas to hold surfaced logs, others still hold logs in weedy areas rife with horsetail and other serious landscape pests. The mills then separate what they call "logyard trash" (a mixture of soil, rock, bark, and fine organic matter) into useful portions. The fines will often have been contaminated with seeds from weeds in the log yard. I have seen a number of landscaped sites where applied bark mulch immediately gave birth to horsetail seedlings. Similar problems have been reported in agricultural studies where bark mulch was used in areas of fruit production.

Third, bark mulch made from trees that have been held in salt water can contain extremely high salt levels, leading to plant stress and death. The curators at the Weyerhaeuser Pacific Rim Bonsai Collection once had the misfortune of using salt-contaminated bark mulch on their trees, causing a significant amount of damage to these valuable specimens. This is particularly a problem with bark obtained from lumber mills around bodies of salt water, where logs are often held before processing. Fortunately, rainwater will eventually leach these salts away.

Finally, bark mulch made from softwood, like Douglas fir, can be miserable to work with! The long, pointed fibers that help support the living cells in the tree are made from the same hard material you find in nutshells. Anyone who has worked in a landscape mulched with bark has probably experienced "porcupine hands" afterward. Gloves can minimize this, of course, but some tasks that require fine coordination do call for bare hands.

Obviously aesthetics play a part in choosing a mulch, but when aesthetics interfere with plant health, they are of secondary importance. To me, the drawbacks associated with bark mulch and sawdust do not justify their use on ornamental landscapes. Conversely, arborist wood chips appear to have all of the benefits and none of the problems associated with bark mulch or sawdust. If you dislike the appearance of wood chips, you can ask for more finely chipped material or purchase your own chipper and do it yourself.

The Bottom Line

- Bark contains natural waxes that prevent absorption and release of water in landscapes.
- Bark mulch can be contaminated with weed seeds or salt.
- Sawdust is too fine a material to use as a landscape mulch and will prevent water and gas movement as it compacts.
- Softwood bark mulches are often not "gardener friendly" due to the presence of tiny, sharp fibers.
- Arborist wood chips can be finely chipped, if this is more aesthetically desirable.

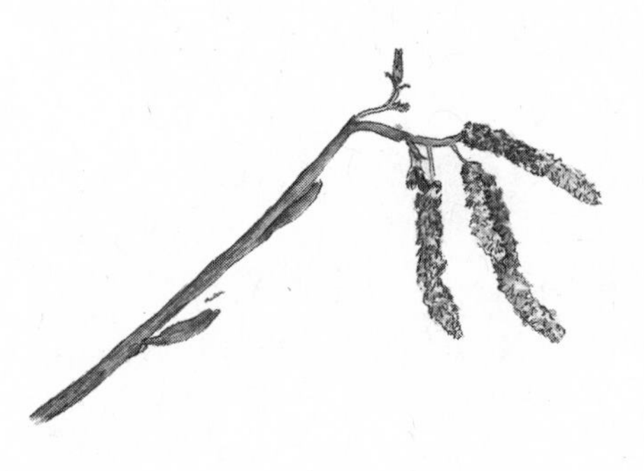

References

Gut, W., F. Weibel, and W. Jaggi. 1990. "Effects of mulching materials on soil fertility." *Schweizerische Zeitschrift fur Obst und Weinbau* 126:685–91.

Stenn, H. 2005. Woody mulch research review, professional users and product availability surveys. Seattle, WA: Seattle Public Utilities.

Original article posted in July 2002.

CHARACTERISTICS OF AN IDEAL LANDSCAPE MULCH

- reduces erosion and compaction
- enhances water infiltration and retention
- enhances gas transfer
- moderates soil temperatures
- improves soil structure
- provides mineral nutrients
- neutralizes pollutants
- enhances beneficial microbes and insects
- suppresses pathogens and pests, including weeds
- is cheap, readily available, and easy to apply
- is aesthetically pleasing

Almost any material will help prevent soil erosion and compaction. Chunky mulch materials are more permeable to water and gas transfer than fine-textured and sheet-type mulches. Organic mulches are generally superior in their ability to supply nutrients, enhance beneficial microbes and nutrients, and suppress pests and disease.

MIRACLES IN A BAG/BOTTLE/BOX

THE MYTH OF COMPOST TEA

The Myth

"Compost tea is an effective alternative to traditional pesticides."

Well, maybe it's too early to call this one a myth. Let's just say (to paraphrase Mark Twain) that news of its effectiveness has been greatly exaggerated. What troubles me is the speed at which this claim has been accepted as a demonstrable fact, when in truth there are only a handful of peer-reviewed publications on compost teas or extracts. There are, however, a number of articles published in popular magazines (such as *Biocycle*), but such articles are not subject to peer review and are considered to be "gray literature" by the scientific community.

Compost teas and extracts have traditionally been used as liquid organic fertilizers, but recently have been touted as powerful antimicrobial agents capable of combating pathogens associated with foliar and fruit diseases. Anecdotal evidence abounds, but controlled, replicable experiments do not. A quick search on the Internet revealed that most of the Web sites containing the phrase "compost tea" are .com sites: Most are selling something. The few .edu sites that do exist are cautious in regard to the miraculous properties associated with compost teas.

The Reality

One of the biggest problems with compost, and by extension teas and extracts, is the high variability among composts from different sources as well as from different batches. Before we can attribute any benefit to a specific compost or compost tea, we must know exactly what materials are in the compost. Furthermore, we need to determine the chemical properties of the compost (i.e., percent of nitrogen, pH, et cetera). And, most important, we must understand what comprise the active ingredients of the compost—are they beneficial microbes? Or are they allelopathic compounds—naturally occurring chemicals with pesticidal qualities?

With the variability that exists among batches of compost tea, it is difficult to interpret results in any scientifically meaningful way. Some compost teas apparently contain large numbers of beneficial microbes that compete for space on leaves and fruits, denying pathogens space to colonize. Others apparently contain antimicrobial chemical compounds produced through decomposition that inhibit pathogen growth. But in the peer-reviewed literature, the only article I was able to find on field-tested compost tea reported no difference in disease control between compost tea and water.

We recently conducted research on the effectiveness of compost tea in alleviating cherry blossom brown rot at the Washington Park Arboretum in Seattle. At the end of the experiment, we will be able to report results that will either support or reject the hypothesis that our specific compost tea prevents cherry blossom brown rot. Although our study is not characterizing the compost tea (e.g., what its chemical properties are), we will be able to recommend what direction future research should take.

An article advocating compost tea usage in *University Week* (University of Washington's weekly paper) suggested that we (at the University of Washington) "need to be more cutting edge with our horticultural practices." As an academic and a horticultural scientist, I agree completely, but practices need to be validated through the scientific process before they can be recommended.

In addition to the fuzzy science that accompanies compost tea usage, I am very concerned with the potential high-nutrient load when such teas are used as fertilizer. Unlike compost used for mulch, which provides a slow release of nutrients, compost teas increase the levels of nitrogen, potassium, and other minerals all at once. It is unlikely that these are completely absorbed by the plants and instead may contribute to the eutrophication of watersheds.

The Bottom Line

- Compost teas have not been suitably characterized, nor have their purported benefits been validated scientifically.

- Compost teas can be overused and potentially contribute to ground-water pollution.

- Properly composted organic material makes a wonderful mulch.

References

As of 2007, there are thirty-four papers addressing anaerobic compost tea effects on disease control, compared to seven papers on aerobic compost tea effects. For the sake of brevity, I'm including a very limited number of the most recent publications on this topic.

Literature Reviews

Litterick, A. M., L. Harrier, P. Wallace, C. A. Watson, and M. Wood. 2004. "The role of uncomposted materials, composts, manures, and compost extracts in reducing pest and disease incidence and severity in sustainable temperate agricultural and horticultural crop production—A review." *Critical Reviews in Plant Sciences* 23(6): 453–79.

Scheuerell, S., and W. Mahaffee. 2002. "Compost tea: Principles and prospects for plant disease control." *Compost Science and Utilization* 10(4): 313–38.

Anaerobic Compost Tea and Disease Control

Al-Dahmani, J. H., P. A. Abbasi, S. A. Miller, and H. A. J. Hoitink. 2003. "Suppression of bacterial spot of tomato with foliar sprays of compost extracts under greenhouse and field conditions." *Plant Disease* 87(8): 913–19.

El-Masry, M. H., A. I. Khalil, M. S. Hassouna, and H. A. H. Ibrahim. 2002. "*In situ* and *in vitro* suppressive effect of agricultural composts and their water extracts on some phytopathogenic fungi." *World Journal of Microbiology and Biotechnology* 18(6): 551–58.

Mello, A. F. S., S. de A. Lourenco, and L. Amorim. 2005. "Alternative products in the in vitro inhibition of *Sclerotinia sclerotiorum*." *Scientia Agricola* 62(2): 179–83.

Ozer, N., and N. D. Koycu. 2006. "The ability of plant compost leachates to control black mold (*Aspergillus niger*) and to induce the accumulation of antifungal compounds in onion following seed treatment." *BioControl* 51(2): 229–43.

van Os, G. J., and J. H. van Ginkel. 2001. "Suppression of *Pythium* root rot in bulbous *Iris* in relation to biomass and activity of the soil microflora." *Soil Biology and Biochemistry* 33(11): 1447–54.

Wang, P. C., and J. W. Huang. 2000. "Characteristics for inhibition of cucumber damping-off by spent forest mushroom compost." *Plant Pathology Bulletin* 9(4): 137–44.

Welke S. E. 2005. "The effect of compost extract on the yield of strawberries and the severity of *Botrytis cinerea*." *Journal of Sustainable Agriculture* 25(1): 57–68.

Aerobic Compost Tea and Disease Control

Al-Mughrabi, K. I. 2006. "Antibiosis ability of aerobic compost tea against foliar and tuber potato diseases." *Biotechnology* 5(1): 69–74.

Diánez, F., I. Trillas, M. Avilés, J. C. Tello, M. Santos, A. Boix, and M. de Cara. 2006. "Grape marc compost tea suppressiveness to plant pathogenic fungi: Role of siderophores." *Compost Science and Utilization* 14(1): 48–53.

Scheuerell, S. J., and W. F. Mahaffee. 2004. "Compost tea as a container medium drench for suppressing seedling damping-off caused by *Pythium ultimum*." *Phytopathology* 94(11): 1156–63.

Sturz, A. V., D. H. Lynch, R. C. Martin, and A. M. Driscoll. 2006. "Influence of compost tea, powdered kelp, and Manzate Reg. 75 on bacterial-community composition, and antibiosis against *Phytophthora infestans* in the potato phylloplane." *Canadian Journal of Plant Pathology* 28(1): 52–62.

Utkhede, R., and C. Koch. 2004. "Biological treatments to control bacterial canker of greenhouse tomatoes." *Biocontrol* 49(3): 305–13.

Compost Tea Problems

Duffy, B., C. Sarreal, S. Ravva, and L. Stanker. 2004. "Effect of molasses on regrowth of *E. coli* O157: H7 and *Salmonella* in compost teas." *Compost Science and Utilization* 12(1): 93–96.

Kannangara, T., T. Forge, and B. Dang. 2006. "Effects of aeration, molasses, kelp, compost type, and carrot juice on the growth of *Escherichia coli* in compost teas." *Compost Science and Utilization* 14(1): 40–47.

Original article posted in April 2001.

THE MYTH OF COMPOST TEA REVISITED

The Myth

"Aerobically brewed compost tea suppresses disease."

When I first addressed the use of compost tea as a disease suppressor, I was concerned about the lack of scientific data documenting the success of compost teas, especially aerobic teas, in disease control. Since then, the popular press and the Internet have exploded with kudos for aerated compost tea as a disease-control agent. There are well over 4,000 .com hits on the Google search engine, compared with only 1,900 two years ago. Numerous magazine and newspaper articles have featured compost teas as environmentally friendly alternatives to chemical pesticides, claiming reduced runoff into aquatic systems among other benefits. As this topic continues to generate more inquiries than any of my other columns, I thought it was time to look at the literature to see what's been added and also to summarize the results of our pilot study.

The Reality

Once again I searched the scientific literature using various combinations of the words "compost," "tea," "leachate," "extract," and "disease." The search engines I use contains all of the life-science-related databases, such as Agricola, Water Resources, Biosis, et cetera.

I limited my review to scientific journals and books published through scientific organizations and academic publishers. I excluded the thirteen articles in *Biocycle*, *Arbor Age*, and *IPM Practitioner*; they, like other trade journals, are not scientifically reviewed. There were published abstracts from scientific meetings which I did not include for the same reason. Peer review is critical to the scientific community, as it allows other researchers in the field to examine manuscripts before they are published. When an article appears in a peer-reviewed journal, it means the methods, results, and conclusions were found to be scientifically viable by objective outside scientists.

I found eighteen articles discussing the disease-suppressing properties of **composts** (especially those containing bark). Researchers have found them effective in suppressing soil diseases such as *Colletotrichum orbiculare* (anthracnose), *Fusarium oxysporum* (wilt) and *F. solani*, *Phytophthora cinnamomi* and *P. cactorum*, *Plasmodiophora brassicae* (clubroot), *Pseudomonas syringae*, *Pythium ultimum* and *P. aphanidermatum* (damping-off disease), *Sclerotinia minor*, *Sclerotium rolfsii* (southern blight), *Sepedonium* species and *Verticillium fungicola*. Beneficial microorganisms colonizing the compost, pathogen inhibition by chemicals passively leaching through the compost, and reduced rainwater dispersal of pathogens from mulch compared to bare soil were identified respectively as the biological, chemical, and physical mechanisms responsible for disease suppression. (Just searching for "compost" and "disease suppression" through these databases netted 100 or so articles. It's clear that compost used as mulch has documented abilities to suppress soilborne disease organisms.)

Over a dozen scientific articles looked at the effects of **non-aerated compost extracts.** (It should be mentioned that the use of non-aerated, or anaerobic, compost teas does not cause your plant or landscape to become anaerobic; anaerobic soil environments occur due to poor drainage, overwatering, soil compaction, or high clay content.) Work in Germany in the 1980s reported success in controlling *Plasmopara viticola*, *Uncinula necator*, and *Pseudopeziza tracheiphila* on grape; *Phytophthora infestans* on potato and tomato; *Erysiphe* species on barley and sugar beet; *Sphaerotheca fuliginea* on cucumber; and *Botrytis cinerea* on strawberry and bean, but these results have apparently not been repeated elsewhere. (Unfortunately, much of this work was published as annual reports, which are not reviewed by the scientific community. In contrast, a later paper found no significant reduction in *Plasmopara viticola* in grape after treatment with compost extracts.) Four articles by another lab group discuss the inhibition of disease using spent mushroom compost extract; prevention of *Venturia inaequalis* (apple scab) was successful in the lab and had mixed success the field. Still another group used leachates from composted bark to inhibit five *Phytophthora* species in the lab with variable success. A final paper reports that compost extracts were variable in their effectiveness in reducing *Colletotrichum orbiculare* in cucumber and *Pseudomonas syringae* in *Arabidopsis*.

In my search, I found only two published, controlled studies using **aerated compost teas**—composts that are extracted in continuously aerated water. One paper reported that aeration of non-aerated compost teas decreased their efficacy in controlling pathogens, but efficacy could be recovered if the teas were allowed to incubate and become anaerobic again. The second paper reported that aerated compost tea "was not effective in preventing scab infection and in some cases appeared to enhance apple scab."

Some additional articles raised environmental concerns: Six articles from my search considered the potential of compost leachates to contribute to water pollution through the excessive release of nitrogen, phosphorus, and other nutrients. This continues to be an area of concern, especially when teas are used as fertilizers. Any overapplication of chemicals, whether from synthetic or natural sources, is potentially harmful to terrestrial and aquatic ecosystems. To assume that "natural" means "safe" is erroneous and environmentally irresponsible.

Why are there so few published data on aerated compost tea? Part of the reason is that each microbe in compost tea needs to be isolated and identified, then tested in a scientifically acceptable method, before it is deemed an effective colonizer and competitor. Each batch of compost tea contains a variety of microbes at varying concentrations. How do these microbes interact? Do combinations of microbes have the same, better, or worse effects than those isolated examples? The potential for variability is enormous, and this leads to inconclusive results during testing.

We experienced this frustration ourselves. In the spring of 2001 we compared the efficacy of compost tea in controlling cherry blossom brown rot (*Monilinia fructicola*) on matched pairs of eight *Prunus* species at the Washington Park Arboretum. Phil Renfrow of the City of Seattle's Department of Parks and Recreation brewed our compost tea in a Growing Solutions twelve-gallon microbe brewer. He used a high-quality aerobic compost that was analyzed and approved by Soil Food Web laboratories. Every week fresh tea was sprayed on half the trees; the others were sprayed with water. Brown rot damage to the blossoms was ranked on a scale from one (no damage) to five (complete infection). In comparing the *Prunus* trees, the effect of the compost tea extract was not significantly different from that of the water application for any of the eight cultivars tested. In fact, for some trees the compost tea made the problem worse. All of these trees have since been removed from the arboretum, so we were not able to repeat this study. However, our results are very much in agreement with another unpublished study in Massachusetts, which found no difference between compost tea and water in preventing *Alternaria* blight or *Septoria* leaf spot in tomatoes.

I have a home landscape with many trees, shrubs, and groundcovers. I don't use pesticides, except for an occasional shot of Roundup; I don't use fertilizers, unless I can determine a deficiency (most commonly nitrogen, which I add as fish meal only to plants that need it); I don't add anything else to the landscape except wood chips as an organic mulch. I don't have disease problems; I don't have insect pests; I have a healthy, organic landscape. This tells me that compost tea is not crucial for landscape health. If a landscape has serious soil or plant-health problems, it is not likely that compost tea is going to solve the problem. Often in urban areas the landscape problems are soil compaction, overuse of fertilizers (especially phosphate), and overuse of pesticides (especially fungicides that harm soil health) Poor plant quality, improper plant siting and installation, and lack of proper aftercare also increase plant-health problems. Adding compost tea will not solve these problems.

The Bottom Line

- Non-aerated compost teas may be useful in suppressing some pathogens on some plants.
- Aerated compost teas have no scientifically documented effect as pathogen suppressors.
- Overuse and runoff of compost teas could conceivably contribute to water pollution.
- There is no "silver bullet" for plant-health problems caused by poor soil health and improper plant selection and management.
- Composted mulch has been documented to suppress disease through a variety of methods.

References

See references in "The Myth of Compost Tea."

Original article posted in August 2003.

THE MYTH OF MINERAL MAGIC

The Myth

"Adding potassium or magnesium to your landscape plants will increase their cold hardiness."

A recent article in a popular gardening resource contained several suggestions for gardeners who like to push the "hardiness envelope" and grow landscape plants outside their natural climate. This information is generally useful for protecting marginally cold-hardy species, though one suggestion was new to me. An addition of wood ashes (for potassium) and Epsom salts (for magnesium) was recommended to promote the cold hardiness of plants grown outside their range. Further investigation on the Web revealed this to be common knowledge among both popular and some educational sources. In particular, potassium is mentioned hundreds of times as "promoting cold hardiness, disease resistance, and general durability." We are informed by one horticultural consultant that "what the potassium does is strengthen the cell walls of the plant and will displace the amount of water in the cells making it harder for them to burst during freezing temperatures." The recommendation to add magnesium for increased cold hardiness is less common on the Web, and the rationale behind its use is not clear. Where did these recommendations originate?

The Reality

Before we address the science behind potassium and magnesium use in improving cold hardiness, I first need to dispel the notion that freezing temperatures cause plant cells to burst. This is a common misconception, but in fact rarely happens in nature. What generally happens is that water freezes in the spaces between the cell walls, and this ice formation draws liquid water from the living cells. The living cells are stressed and sometimes killed by the dehydration caused by water moving into these spaces. The frozen water in these spaces does not injure the plant, since it is outside the cell membranes. Rapid freezing rates will cause water inside the cells to freeze and rupture cell membranes—but not cell walls.

There are a number of scientific publications that address the involvement of potassium and/or magnesium in cold hardiness development. Briefly, potassium is vitally important in regulating cell membrane activity and water relations within the plant. A deficiency in potassium results in foliar symptoms including chlorosis, necrosis, warping, and cupping. If a plant is deficient in potassium, adding this macronutrient will obviously help. Potassium levels can be deficient in acid sandy soils and in intensively managed farmland, but is rarely limiting in nonagricultural soils.

While potassium's importance in water relations is clear, its role in cold hardiness is anything but. In eight separate studies on coniferous species, additional potassium had no effect on cold hardening, while four other studies found a negative relationship between increased potassium and improved cold hardiness. Two studies found an improvement in cold hardiness: one with actively growing turf grass, and one with hydroponic seedlings. Three studies found that additional potassium is related to spring hardiness (i.e., late frosts) of vegetative and floral buds that have already broken dormancy. There is absolutely no scientific evidence, however, that additional potassium will increase the winter hardiness of marginally hardy landscape species.

Magnesium, like many metals, assists in enzyme activity and, like potassium, is generally not limiting in nonagricultural soils unless they are quite alkaline. Two conflicting papers looked specifically at interactions between cold hardiness and magnesium in coniferous species. In one, cold damage is associated with high soil magnesium. The other paper reports the opposite relationship. It is not clear why magnesium has been associated with cold hardiness; it certainly is not based on available science.

There are about a dozen other papers that considered potassium and magnesium either together or in conjunction with other minerals. Some of the studies' conclusions conflicted with others; cold hardiness is reported to increase, decrease, or not change as a result of the nutritional regime in question. Potassium and magnesium were occasionally reported to interfere with each other when added in excess. The combined results of these studies can be easily summed up in the words of one of the authors: "There was no clear relation between the pattern of frost hardiness and nutrient concentrations."

Many of these papers recognize that cold hardiness is not

affected by the presence or absence of one mineral nutrient but instead is influenced by other environmental conditions. Abiotic factors—including drought, flooding, compaction, salinity, mineral deficiencies and toxicities, and air pollutants such as ozone and acid deposition—can all increase or decrease cold hardiness, as do biotic factors—including pests, disease, and human activities. Often, mild environmental stresses will increase the plant's resistance to that particular stress as well as others, but at other times stresses will make the plant more susceptible to further stress. There is no simple model for predicting environmental stress interactions, and certainly no magic bullet for preventing or treating the plants exposed to them.

It is disappointing that so many university Extension Web sites promote the myth of mineral magic. While a few species may tend toward potassium deficiency (as do certain turf grasses and palms, for example), this condition cannot be generalized to include all woody trees and shrubs. Even where deficiencies exist, there is no rational linkage between deficiency and cold hardiness. A few Extension Web pages do report the lack of science behind this myth, but they are unfortunately the exceptions.

In brief, the best way to prevent cold damage to marginally hardy plant material is to install plants where the microclimate is warmest year round (e.g., leeward of winter winds) and to insulate both the plant and the soil surface from cold temperatures (e.g., by using mulch). During the growing season, be sure that other environmental factors are optimal; a healthy plant has a better chance of surviving winter stress. In the winter, the critical thing is to prevent the plant from experiencing lethal low temperatures, and no amount of any mineral element will accomplish this.

The Bottom Line

- There is little evidence that the addition of either potassium or magnesium will increase the hardiness of native or of nonnative, marginally hardy landscape species.

- To grow marginally hardy species, take advantage of microclimates to maximize their chances for survival.

- The best strategy for overwintering marginally hardy species is to insulate them and the surrounding soil.

- Before adding any mineral supplement to your landscape, have a soil test done first to determine if deficiencies exist.

- The addition of chemicals (organic or inorganic) to a landscape where no mineral deficiency exists is a waste of money, time, and resources, and is environmentally irresponsible.

References

Bigras, F. J., J. A. Rious, R. Paquin, and H. P. Therrien. 1989. "Effect of extending fertilizer application into the autumn on frost tolerance and spring growth of container-grown *Juniperus chinesis* 'Pfitzerana.'" *Phytoprotection* 70(2): 75–84.

Birchler, T. M., R. Rose, and D. L. Haase. 2001. "Fall fertilization with N and K: Effects on Douglas-fir seedling quality and performance." *Western Journal of Applied Forestry* 16(2): 71–79.

Edwards, G. S., P. A. Pier, and J. M. Kelly. 1990. "Influence of ozone and soil magnesium status on the cold hardiness of loblolly pine (*Pinus taeda* L.) seedlings." *New Phytologist* 115(1): 157–64.

Hawkins, B. J., G. Henry, and J. Whittington. 1996. "Frost hardiness of *Thuja plicata* and *Pseudotsuga menziesii* seedlings when nutrient supply varies with season." *Canadian Journal of Forest Research* 26(8): 1509–13.

Jalkanen, R. E., D. B. Redfern, and L. J. Sheppard. 1998. "Nutrient deficits increase frost hardiness in Sitka spruce (*Picea sitchensis*) needles." *Forest Ecology and Management* 107(1–3): 191–200.

Jokela, A., T. Sarjala, and S. Huttunen. 1998. "The structure and hardening status of Scots pine needles at different potassium availability levels." *Trees: Structure and Function* 12(8): 490–98.

Miller, G. L., and R. Dickens. 1996. "Potassium fertilization related to cold resistance in bermudagrass." *Crop Science* 36(5): 1290–95.

Sarjala, T., K. Taulavuori, E. M. Savonen, and A. B. Edfast. 1997. "Does availability of potassium affect cold hardening of Scots pine through polyamine metabolism?" *Physiologia Plantarum* 99(1): 56–62.

South, D. B., D. G. M. Donald, and J. L. Rakestraw. 1993. "Effect of nursery culture and bud status on freeze injury to *Pinus taeda* and *P. elliottii* seedlings." *South African Forestry Journal* 166:37–45.

Original article posted in October 2004.

THE MYTH OF VITAMIN SHOTS

The Myth

"Fertilizer injection is the most effective way to correct tree nutrient deficiencies."

One of the more interesting members of the landscape management community is the medical professional. Advertisements show white-coated technicians with the latest injection equipment for delivering doses of fertilizers and fungicides to ailing plant patients. The memory of the family doctor who would make house calls is embodied in these plant-health care professionals, and we feel we are providing the best possible care for our shrubs and trees.

A great deal of scientific literature has been dedicated to this practice. Leaf chlorosis (yellowing) tends to be the primary signal to landscape managers that fertilizer application is required. Fertilizer has been injected into the surrounding soil and trunks of a number of tree species, generally those with landscape or fruit and nut production value. In addition to complete fertilizers, specific nutrients such as iron, zinc, and magnesium have also been injected. Proponents of injection fertilization, especially trunk injection, point to the immediate improvement in leaf color and the relatively low cost of application as a justification for the practice.

The Reality

This is another practice whose immediate results should be weighed against the long-term effects. Indeed, trunk injection of nutrients can have an immediate impact on leaf color, but what is routinely missing in published reports and papers is the long-term effect of the practice on leaf color and tree health. Indeed, when studies are carried through for a number of seasons, invariably the authors will report that the leaf color of the injected trees is no different from that of the control trees. This is not a practice with sustainable benefits.

Injecting plants with various substances has fascinated humans for many centuries. The fertilization practices of tree injection and soil injection have been studied for decades and have been recently reviewed in the scientific literature. It is apparent in these reviews that there are no long-term benefits to trunk injection of fertilizer. Though it can have immediate effects in terms of leaf color change, trunk injection breaches the tree's bark barrier and leads to numerous health problems. Injection sites are portals for pathogens and pests; they can cause trunk splitting, decay, cankers, and structural defects; and they are especially dangerous to trees already in poor condition. Furthermore, trees injected with fertilizer have been found to become more susceptible to insect pests, presumably because their leaf nitrogen content increases.

Oddly enough, this practice has been repudiated several times throughout the twentieth century by agricultural and silvicultural researchers. Unfortunately, it continues to be widely recommended and practiced, since it "makes sense" to people who assume that medical and veterinary models extend to the care of plants. Though the value of vaccines and other medical injection procedures for animals (including humans) are clear and documented, it is not a practice that translates to plant species, whose normal physiological functions and biochemical resistance strategies are still poorly understood.

What about soil injection? There are several papers and a few scientific reviews on soil injection of fertilizers, though not nearly as many as on trunk injection. Briefly, researchers have studied the application of nutrients such as nitrogen, potassium, and iron to various landscape and crop trees. Though not an invasive technique like trunk injection, soil injection does not significantly improve the delivery of fertilizer to tree roots. In fact, a review in 2002 concluded that "surface applications were as effective as soil injection or drilling." This appears to be a practice that generally is not warranted and adds excessive costs to maintaining a landscape.

We need to be more aware of mitigating factors that affect leaf color and overall tree health. Competition from weeds and turf will reduce the availability of nutrients to trees and shrubs, whose roots are often deeper in the soil. Poor soil conditions (including compaction and waterlogging, lack of adequate irrigation and mulch, opportunistic pests and pathogens, weeds, and urban stresses) can all contribute to tree decline. Yellowing leaves do not necessarily mean that fertilizer is necessary; they are often a sign of other plant stresses that no amount of fertilizer will correct (though weeds and turf will readily respond). A proactive approach to mineral nutrition of trees, especially using soil and leaf tissue analyses to determine what elements are truly deficient, is more economically and environmentally sustainable than the "quick fix."

The Bottom Line

- Trunk injection is harmful to the long-term health of the tree and should not be used for delivering fertilizers.
- Soil injection is no more effective at delivering fertilizer than broadcast application and is not cost-effective.
- Fertilizer will not cure nutritional deficiencies caused by disease, pests, air pollution, mineral toxicity, drought, poor root health, or poor soil health.
- Leaf yellowing and other foliage symptoms are not direct indicators of soil nutrient levels.
- Before any fertilizer program is initiated, a complete soil analysis should be performed.
- Leaf tissue analysis should be performed before initiating any specialized fertilization program.
- Any fertilizer program that does not include weed management is ineffective and costly.
- Woody plants growing in competition with turf will always

suffer more nutrient stress than those trees and shrubs partially protected by mulch.

References

Costinis, A. C. 1980. "The wounding effects of Mauget and Creative Sales injections." *Journal of Arboriculture* 6(8): 204–8.

Hurley, A. K., R. H. Walser, and T. D. Davis. 1986. "Net photosynthensis and chlorophyll content in silver maple after trunk injection of ferrous sulfate." *Journal of Plant Nutrition* 9(3–7): 683–93.

Mayhead, G. J., and D. Bole. 1994. "Treatment of severely checked Sitka spruce trees with injected nutrients." *Forestry* 67(4): 343–49.

McClure, M. S. 1992. "Effects of implanted and injected pesticides and fertilizers on the survival of *Adelges tsugae* (Homoptera: Adelgidae) and on the growth of *Tsuga canadensis*." *Journal of Economic Entomology* 85(2): 468–72.

Perry, T. O., F. S. Santamour Jr., R. J. Stipes, T. Shear, and A. L. Shigo. 1991. "Exploring alternatives to tree injection." *Journal of Arboriculture* 17(8): 217–26.

Raese, J. T., C. L. Parish, and D. C. Staiff. 1986. "Nutrition of apple and pear trees with foliar sprays, trunk injections or soil applications of iron compounds." *Journal of Plant Nutrition* 9(3–7): 987–99.

Smiley, E. T., and L. G. King. 1998. "Longevity of ferric ammonium citrate treatments in oak." *Journal of Arboriculture* 24(5): 294–96.

Struve, D. K. 2002. "A review of shade tree nitrogen fertilization research in the United States." *Journal of Arboriculture* 28(6): 252–63.

Worley, R. E., and R. H. Littrell. 1980. "Ineffectiveness of foliar application and pressure trunk injection of Mg-N formulation for correction of Mg deficiency of pecan." *HortScience* 15(2): 181–82.

Original article posted in December 2003.

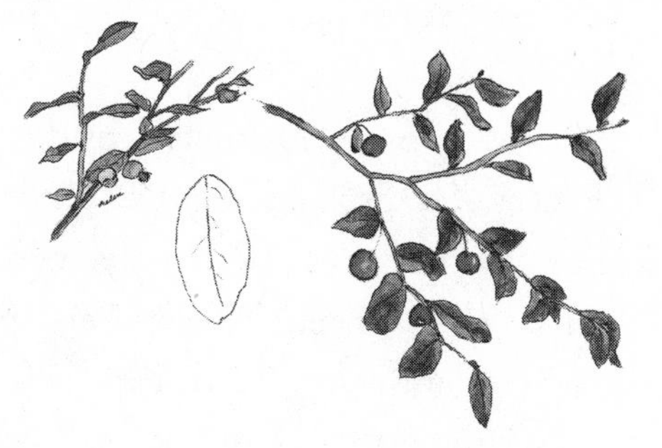

THE MYTH OF VITAMIN STIMULANTS

The Myth

"Vitamin B1 reduces transplant shock by stimulating new root growth."

Ever seen this advertisement? "[Product X, which contains vitamin B1] stimulates the quick formation of new root hairs and revitalizes the delicate feeder roots that are often damaged in transplanting. [Product X] is especially designed to hasten the development of bare-root roses, shrubs, shade trees and bedding plants that have been moved to new locations. It helps plants become established quickly and ensures vigorous growth." Another advertisement adds a little scientific terminology to convince you: "Vitamin B1 (plus minor elements and chelating agents) is great for root growth and helps reduce transplant shock." Or how about this one? "The combination of vitamin B1 with essential micro nutrients forms a highly effective mixture . . . and lessens the chances of transplant shock and plant stress."

Aren't you convinced that if you don't use products with vitamin B1, your transplants will suffer? Apparently administrators at one large university are. In their guide "Typical Tree Protection and Relocation Specifications" is the following: "48 hours prior to cutting, an application of vitamin B1 shall be administered to the rootball of the tree." If a university requires this practice, it must be legitimate, right?

The Reality

Applying vitamin B1, or thiamine, to root systems of whole plants does not stimulate root growth. This is a myth that refuses to die, though it has been repeatedly refuted in the scientific literature. To understand why, it helps to think about this in a historical perspective.

Many decades ago, the plant growth regulators called "auxins" were isolated and characterized. Auxins were found to stimulate cell elongation in both root and shoot tissues. Commercial preparations were developed that contained auxin and vitamin B1, among other ingredients. Research in 1949 found improved root development in plants treated with one of these preparations (Transplantone), but noted the importance of auxins in this response. Further research throughout the last half of the twentieth century investigating the application of auxins to root systems suggested that auxins may stimulate root growth, but that vitamin B1 on its own does not.

So what does stimulate root growth and reduce transplant shock? A review of the historical and current literature reveals the following:

- Indole butyric acid is one of the most common auxin formulations, especially in tissue culture. In cuttings, it has been found to increase the number of roots, to increase rooting percentage, to increase both parameters, or to increase neither. Indole butyric acid has had some success in regenerating the roots of transplanted trees; it may help redirect resources to the roots by suppressing crown growth.

- Naphthylacetic acid is also a commonly used auxin and often the active ingredient in commercial preparations. Naphthylacetic acid tends to be toxic to seedling root development, as it inhibits primary root growth and enhances lateral root growth. This latter activity may account for naphthylacetic acid's success in regenerating the roots of transplanted and root-pruned trees. Like indole butyric acid, naphthylacetic acid apparently suppresses crown growth, which may also help redirect resources to the roots.

- Paclobutrazol is another plant-growth regulator that seems to stimulate root growth in containerized as well as established tree species. Like the auxins, paclobutrazol reduces crown growth, which may assist root resources.

- Fungicides may increase root growth, but overall this is not beneficial to the plant. Fungicides kill beneficial mycorrhizal species, and the lack of mycorrhizal colonization means that plants must put more resources into root growth than they would if mycorrhizae were present. Furthermore, there are beneficial fungi and bacteria that control pathogenic microbes, and roots colonized by beneficial microbes have been shown to grow more than those without.

- Nitrogen supplements can improve root growth, and, conversely, the absence of nitrogen will depress root growth. The competition for nitrogen uptake among bacteria, fungi, and plants can be intense, so nitrogen is often limiting.

- Vitamin B1 is an important component of tissue culture media, in which isolated plant tissues can be propagated. Its use in stimulating root growth in whole plants is not supported in the literature, and one study reported that root growth was greater with the control treatment (water) than with vitamin B1. Plants in the field manufacture their own source of vitamin B1, and it is therefore unnecessary to supplement it. Many fungi and bacteria associated with plant roots also produce vitamin B1, so it's likely that healthy soils will contain adequate levels of this vitamin without amendment.

The mystique of vitamin B1 transplant tonics still persists after decades of scientific debunking. People want to believe in miracles and magic bullets, but in reality, the best transplant tonic for landscape plants is pure, abundant water.

The Bottom Line

- Vitamin B1 does not reduce transplant shock or stimulate new root growth in plants outside the laboratory.

- A nitrogen fertilizer is adequate for transplanting landscape plants; avoid use of "transplant fertilizers" that contain phosphate.

- Difficult-to-transplant species may be aided by the application of auxin-containing products in addition to nitrogen, but read the label and don't add unnecessary and potentially harmful chemicals (this includes organics!).

- Healthy plants will synthesize their own vitamin B1 supply.

- Healthy soils contain beneficial microbes that synthesize vitamin B1 as well.

- Adequate soil moisture is crucial for new root growth; be sure to irrigate new transplants frequently and use mulch to reduce evaporation.

References

Frampton, L. J. Jr., B. Godfarb, S. E. Surles, and C. C. Lambeth. 1999. "Nursery rooting and growth of loblolly pine cuttings: Effects of rooting solution and full-sib family." *Southern Journal of Applied Forestry* 23(2): 108–16.

Kelting, M., J. R. Harris, and J. Fanelli. 1998. "Humate-based biostimulants affect early post-transplant root growth and sapflow of balled and burlapped red maple." *HortScience* 33(2): 342–44.

Migliaccio, C. P. 2000. "The effects of vitamin B1 on palm seedling growth." *Palms* 44(3): 114–17.

Original article posted in April 2004.

THE MYTH OF WOUND DRESSINGS

The Myth

"Apply wound dressing after pruning to insure against insect or fungal invasion."

The late Alex Shigo debunked the myth of wound dressing decades ago with his exhaustive field research on tree responses to wounding. But the myth persists, particularly among those with something to sell. A quick look at the Internet revealed the following claims (I've left out brand names, but they're easy to find):

- "A clean, easy, simple way to aid in healing cuts and protecting tree wounds, pruned-edges and graft unions of roses, trees, and shrubs."

- "Insure your trees, shrubs, and vines against decay, insects, and fungi in any kind of weather."

- "An artificial bark for treatment of wounds . . . made of ALL NATURAL biodegradable materials."

- And my personal favorite (advertised as "an extremely durable rust and corrosion inhibitor") lists a multitude of wonderful uses for their product: truck, farm and construction equipment, collision shops; tree-wound dressing; gutter and flashing sealer; electrical box and fittings sealer; concrete and asphalt driveway crack sealer.

More recently, "green" companies have peddled collagen, pectin, hydrogel, and aloe gel as "natural" tree healers. These hucksters claim that "the surface will heal over quickly and insects are repelled by the bitter taste." Not one shred of scientific evidence is ever offered to substantiate these claims.

The Reality

A tree wound dressing is "a petroleum-based product used to cover freshly cut wood to inhibit decay or insect infestation" (from the glossary on Regenesis.net). Yikes! Think about this stuff—a petroleum-based product. Does this sound like a substance that would be beneficial to a living tissue? Would you use it to treat a cut on your own skin? If the idea repels you, carry that logic over to plant health care. (Note: while petroleum jelly is an obvious exception, it also is not used as a tree-wound dressing.)

Tree-wound dressings do, however, have a number of effects,

few of which are desirable from a tree-health perspective. These waterproof substances will seal in moisture and decay, and in some cases serve as a food source for pathogens. Wound dressings prevent the tree's own defense mechanisms from operating well, since oxygen is required for the chemical reactions that form wound wood. This wound wood helps compartmentalize, and thus isolate, wounds from the rest of the tree, limiting the ability of pathogens to invade. Wound dressings not only inhibit this natural process, but eventually crack, exposing the tree to pathogens. Thus, wound dressings do not prevent the entrance of decay organisms, nor will they stop rot once that process has begun.

For some inexplicable reason, people are compelled to "manage" a process that plants have evolved over millions of years. Look at this advice from another Web site with something to sell: "Bark with cracks. This is a *natural* occurrence in the growth of trees & shrubs. Where cracking occurs in the lower wood it is probably caused by flooding after a long period of dryness. Cracks in the trunk should be painted with a tree wound dressing or bituminous paint to prevent the invasion of fungal diseases [emphasis mine]."

It's important to recognize that trees do not "heal." Instead, they isolate damage through formation of suberized, lignified wood that physically and chemically repels invasion. A callus develops at the edge of the wound and gradually expands toward the center. This wound wood remains for the life of the tree; bark does not regenerate itself the same way our skin does. Every year, trees form hundreds of tiny abscission layers as leaves senesce and fall. Wounds left from branch breakage are callused over and compartmentalized.

There may be some benefit in treating the wounds of trees that are particularly susceptible to certain diseases, such as oak wilt. Many regions in the country specify that oaks pruned in areas where oak wilt is a problem should be treated to prevent infection. While research supporting this advice is sketchy at best, it may be justifiable to use a fungicide or insecticide during spring or summer pruning. If pruning is done during the dormant season, the chance of infection is greatly reduced, and wound treatment should be avoided.

Finally, the use of wound dressing "for aesthetic reasons" is never justified. In this case, the customer is not "always right." These situations should serve as opportunities to educate the tree owner.

The Bottom Line

- Like all living organisms, plants have natural resistance mechanisms to fight insect attacks and diseases.

- Covering wounds with traditional sealants inhibits the oxidative processes, which in turn will reduce callus formation and subsequent compartmentalization.

- The optimal pruning time for insect- or disease-prone species is in the fall or winter when temperatures and infection rates are lower.

- If you must prune a disease-prone species when insects or fungi are active (i.e., during the warmer times of the year), a light coating of an insecticide or fungicide may be warranted.

- Try sterilizing pruning tools. Such measures can help reduce the transmission of certain plant diseases to healthy plants.

- Disease spread can be controlled through preventative management practices such as the disposal of contaminated organic material and the use of disease-free compost and mulch.

References

Balder, H., and G. Kruger. 1995. "Investigations of the effect of wound dressing after pollarding in plane trees." *Nachrichtenblatt des Deutschen Pflanzenschutzdienstes* 47(1): 1–4.

Ball, J. 1992. "Response of the bronze birch borer to pruning wounds on paper birch." *Journal of Arboriculture* 18(6): 294–97.

Barger, J. H., and W. N. Cannon Jr. 1987. "Response of smaller European elm bark beetles to pruning wounds on American elm." *Journal of Arboriculture* 13(4): 102–4.

Dujesiefken, D., T. Kowol, and E. Schmitz-Felten. 1996. "The influence of time of wounding on the effectiveness of wound dressings on deciduous trees." *Gesunde Pflanzen* 48(3): 89–94.

English, H. 1958. "Physical and chemical methods of reducing *Phomopsis* canker infection in Kadota fig trees." *Phytopathology* 48:392.

Neely, D. "Healing of wounds on trees." 1970. *Journal of the American Society for Horticultural Science* 95(5): 536–40.

Shigo, A. L., and C. L. Wilson. 1977. "Wound dressings on red maple and American elm: Effectiveness after five years." *Journal of Arboriculture* 3(5): 81–87.

Shigo, A. L., and C. L. Wilson. 1972. "Discoloration associated with wounds one year after application of wound dressings." *Arborist's News* 37(11): 121–24.

Original article posted in October 2000.

ESSENTIAL GARDEN TOOLS AND PRODUCTS

INDISPENSABLE TOOLS

- folding pruning saw
- garden hose
- good quality shovel or transplanting spade
- hand weeder (or weeding hoe)
- heavy-duty rake
- high-quality bypass-style hand pruners (not anvil pruners; these crush plant tissues)
- hori-hori
- knee pads (or bench)
- large plastic tub (for soaking root balls before planting)
- mulch fork
- soaker hoses

NICE TO HAVE

- chipper for creating your own mulch
- gardening gloves
- irrigation timers
- rain barrels
- sharpening system (for pruners and shovels)
- watering can

PRODUCTS

- glyphosate (Roundup)
- high-nitrogen, nonphosphate fertilizer (organic or inorganic; be sure NPK label has "0" as the middle number); use this when transplanting
- insecticidal soap

REFERENCE BOOKS

- *Arboriculture.* 4th ed. 2004. R. W. Harris, J. R. Clark, and N. P. Matheny. Upper Saddle River, NJ: Prentice Hall Inc.
- *Biology of Plants.* 7th ed. 2005. P. H. Raven, R. F. Evert, and S. E. Eichorn. New York, NY: W. H. Freeman.
- *Plants of the Pacific Northwest Coast.* 1994. J. Pojar and A. MacKinnon. Redmond, WA: Lone Pine Publishing.
- *Urban Soil in Landscape Design.* 1992. P. J. Craul. New York, NY: John Wiley & Sons, Inc.

INDEX